Computer Solutions in Physics

With Applications in Astrophysics, Biophysics, Differential Equations, and Engineering

Computer Solutions in Physics

With Applications in Astrophysics, Biophysics, Differential Equations, and Engineering

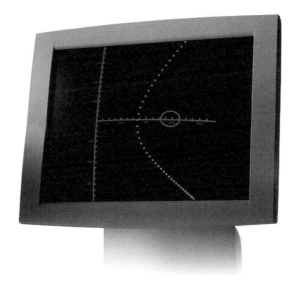

Steve VanWyk

Olympic College, USA

 **World Scientific**

NEW JERSEY · LONDON · SINGAPORE · BEIJING · SHANGHAI · HONG KONG · TAIPEI · CHENNAI

Published by

World Scientific Publishing Co. Pte. Ltd.

5 Toh Tuck Link, Singapore 596224

USA office: 27 Warren Street, Suite 401-402, Hackensack, NJ 07601

UK office: 57 Shelton Street, Covent Garden, London WC2H 9HE

Library of Congress Cataloging-in-Publication Data
Van Wyk, Steve.
    Computer solutions in physics : with applications in astrophysics, biophysics, differential
equations, and engineering / Steve Van Wyk.
       p. cm. + 1 CD-ROM
    Includes bibliographical references and index.
    ISBN-13 978-981-270-936-3
    ISBN-10 981-270-936-3
    ISBN-13 978-981-277-499-6 (pbk)
    ISBN-10 981-277-499-8 (pbk)
    1. Physics--Data processing. I. Title.
    QC52 .V36 2008
    530.0285--dc22
                                                                    2008301002

British Library Cataloguing-in-Publication Data
A catalogue record for this book is available from the British Library.

First published 2008
Reprinted 2010

Printed in Singapore by World Scientific Printers

Introduction

About thirty years ago, David Park wrote a book on Quantum Theory in which he advised the student to determine whether the difficulty in solving a problem was in the Physics or in the Mathematics. With the great amount of progress in numerical methods and with the speed of the modern personal computer, the problem is no longer in the mathematics. Now it is sufficient to set up the correct Physics equations and utilize any of the excellent mathematical softwares to graph and solve the problem.

The computer solutions in this book are primarily written in *Mathematica* because of its reasonably straightforward approach to mathematical problem-solving. It should be noted that the computational powers of *Maple* and *MatLab* are equally good. A comparison of *Mathematica, Maple,* and *MatLab* is given in Appendix A.

The programming with *Mathematica* 5 should present few difficulties. There are 50-plus computer solutions from physics and engineering and 10-plus animations contained on the compact disc. Any of the *Mathematica* programs on the CD may be run with *Mathematica* Player or *Mathematica* 5, 6 or 7 as appropriate, and MatLab programs may be run with any version of MatLab 7. Once a program from the CD is on your computer, you may change the conditions or modify the equations to address a problem of your choosing. A listing of all programs is given in Appendix B. If you have never used *Mathematica* before, a capsule summary of *Mathematica* commands is given in Appendix C, and a summary of MatLab commands is given in Appendix D.

It is important to note that this is not a first-year textbook on Physics. Anyone who wants to program Physics equations should have at least one year of College Physics and a good introduction to the Calculus.

Here is the premise of this text: If you can write down the correct Physics equations, then it is only necessary to program a few lines of code to get the answer. And if the Physics equations are not correct, then the program output will tell you that as well. Either way, you win. Let's get started by setting up some Physics equations ... and let the computer knock them down.

Contents

Chapter 1 Equations of Motion

1.0   Newton's Laws

Some 320 years ago, Isaac Newton wrote down the two basic equations of classical Physics

the basic equation of force $\qquad F = m\dfrac{dv}{dt} = m\,\dot{v}$

and the force of gravity $\qquad m\,\ddot{r} = -\dfrac{mGM}{r^2}$

Let us begin with motion of a planet in a gravity field. We'll use the metric system to describe the motion of the Earth about the Sun.

Cancelling m, $\qquad \ddot{r} = -\dfrac{GM}{r^2} \qquad$ where $M = M_{sun} = 1.989 \times 10^{30}$ kg

$$\text{and} \quad G = 6.673 \times 10^{-11} \frac{m^3}{kg \cdot s^2}$$

Notice that if we wish to measure distance in km and time in hours, then we need only change the units of G

$$G = 6.673 \times 10^{-11} \frac{m^3}{kg \cdot s^2} = 6.673 \times 10^{-20} \frac{km^3}{kg \cdot s^2}$$

$$G = 3600^2 \times 6.673 \times 10^{-20} \frac{km^3}{kg \cdot (hr)^2}$$

G is exactly the same. Only the units are changed. Choose the G you need, and express your distance units in m or km and your time in s or hr, but be consistent throughout your equation.

Now, back to the Earth. If we place the Earth at its *average distance* $r = 149.6 \times 10^6$ km away from the Sun, in a circular orbit with velocity v,

$$\frac{mv^2}{r} = \frac{mGM}{r^2}$$

$$v = \sqrt{\frac{GM}{r}} = 29.786 \text{ km/s}$$

Let's now write the physics equations for *circular motion* of the Earth about the Sun. To begin, we'll use x-y coordinates.

$$\ddot{x} = -\frac{GM}{x^2 + y^2}\cos\theta = -\frac{GM\,x}{(x^2 + y^2)^{3/2}} \qquad x_0 = 149.6 \times 10^6 \text{ km} \qquad \dot{x}_0 = 0$$

$$\ddot{y} = -\frac{GM}{x^2 + y^2}\sin\theta = -\frac{GM\,y}{(x^2 + y^2)^{3/2}} \qquad y_0 = 0 \qquad \dot{y}_0 = 29.786 \text{ km/s}$$

These two equations describe the motion of the Earth for all time, based on the initial conditions. Let's look at the requirements for finding the speed and location of the Earth at all times after $t = 0$. First of all these two equations are complicated and they are not linear. We shall find a solution by using numerical methods (**NDSolve** in *Mathematica*). We will solve from $t = 0$ to $t = 9000$ hours.

Here is how we do it. To get the time in hours, and the distance from the sun in kilometers, take

$$G = 3600^2 \times 6.673 \times 10^{-20} \frac{\text{km}^3}{\text{kg} \cdot (\text{hr})^2} \quad \text{and} \quad M_{sun} = 1.989 \times 10^{30} \text{ kg}$$

Then the x- and y- distance of the Earth from the center of the Sun will be measured in kilometers. Notice that to be consistent we will also have to convert the initial velocity into kilometers per hour.

Here is the *Mathematica* program

```
In[1]:=  G = 3600^2 * 6.673 * 10^-20;  M = 1.989 * 10^30;
         sol = NDSolve[{
            x''[t] == G * M * (-x[t]) / (x[t]^2 + y[t]^2)^1.5,
            y''[t] == G * M * (-y[t]) / (x[t]^2 + y[t]^2)^1.5,
            x[0] == 149.6 * 10^6, y[0] == 0, x'[0] == 0, y'[0] == 29.786 * 3600},
            {x, y}, {t, 0, 9000}]
```

When we press **Shift-Enter**, the *Mathematica* program returns

```
Out[1]=  {{x → InterpolatingFunction[{{0., 9000.}}, <>],
          y → InterpolatingFunction[{{0., 9000.}}, <>]}}
```

x and y are given by an <u>Interpolating Function</u>. We can plot x vs y if we first identify an interpolation function in x, and an interpolation function in y and table these values versus time.

```
InterpFunc1 = x /. sol[[1]]; InterpFunc2 = y /. sol[[1]];
InterpFunc3 = x' /. sol[[1]]; InterpFunc4 = y' /. sol[[1]];
tbl = Table[{InterpFunc1[t], InterpFunc2[t]}, {t, 0, 8600, 200}];
```

Now we are free to plot a series of (x, y) points versus time. Either as a series of dots (**ListPlot**) or as a continuous line (**ParametricPlot**).

```
ListPlot[tbl, AspectRatio → Automatic, Prolog → AbsolutePointSize[3]]
```

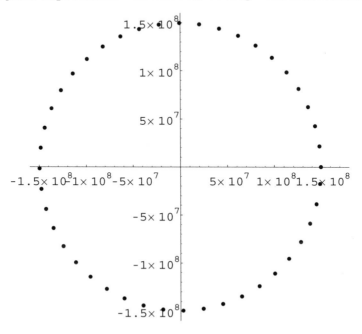

```
ParametricPlot[{x[t], y[t]} /. sol, {t, 0, 7400}, AspectRatio → Automatic]
```

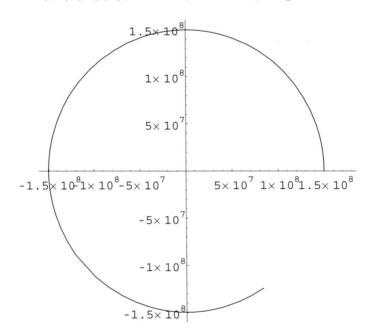

A Parametric Plot of the Earth in a circular orbit.

Let us now **Table** the data from the Interpolation Functions.

```
In[191]:=  r[t_] = √(InterpFunc1[t]^2 + InterpFunc2[t]^2);
           v[t_] = √(InterpFunc3[t]^2 + InterpFunc4[t]^2) / 3600;
           Table[{t, InterpFunc1[t], InterpFunc2[t], r[t], v[t]},
             {t, 0, 8800, 400}] // TableForm
```

t (hr)	x (km)	y (km)	r (km)	v (km / s)
0	1.496×10^8	-1.43599×10^{-21}	1.496×10^8	29.786
400	1.43493×10^8	4.23066×10^7	1.496×10^8	29.786
800	1.25672×10^8	8.11593×10^7	1.496×10^8	29.786
1200	9.75899×10^7	1.13386×10^8	1.496×10^8	29.786
1600	6.15408×10^7	1.36356×10^8	1.496×10^8	29.786
2000	2.04676×10^7	1.48193×10^8	1.496×10^8	29.786
2400	-2.22767×10^7	1.47932×10^8	1.496×10^8	29.786
2800	-6.32023×10^7	1.35594×10^8	1.496×10^8	29.786
3200	-9.8968×10^7	1.12185×10^8	1.496×10^8	29.786
3600	-1.26654×10^8	7.96178×10^7	1.496×10^8	29.786
4000	-1.43999×10^8	4.05503×10^7	1.496×10^8	29.786
4400	-1.49589×10^8	-1.82776×10^6	1.496×10^8	29.786
4800	-1.42966×10^8	-4.40566×10^7	1.496×10^8	29.786
5200	-1.24671×10^8	-8.26886×10^7	1.496×10^8	29.786
5600	-9.61974×10^7	-1.1457×10^8	1.496×10^8	29.786
6000	-5.98704×10^7	-1.37097×10^8	1.496×10^8	29.786
6400	-1.86556×10^7	-1.48432×10^8	1.496×10^8	29.786
6800	2.40823×10^7	-1.47649×10^8	1.496×10^8	29.786
7200	6.48541×10^7	-1.34811×10^8	1.496×10^8	29.786
7600	1.00331×10^8	-1.10968×10^8	1.496×10^8	29.786
8000	1.27617×10^8	-7.80645×10^7	1.496×10^8	29.786
8400	1.44484×10^8	-3.8788×10^7	1.496×10^8	29.786
8800	1.49555×10^8	3.65522×10^6	1.496×10^8	29.786

The above Table shows the Earth in a circular orbit about the Sun.
For our choice of starting conditions, r and v remain constant.

Just how accurate is *Mathematica*? Notice that in the above **Table** that we have solved the Differential Equations for a time of one year, and, as expected, the radius of the orbit and the velocity have remained absolutely constant. This is a first test of the *Mathematica* numerical solver, and it is gratifying to see that the LSODA algorithm in **NDSolve** produces good, consistent results. (The Livermore Solver for Ordinary Differential equations Adaptive method was developed at Lawrence Livermore Labs and utilizes both an Adams method and a Gear backward differences method to obtain results of high accuracy.)

In terms of precision, *Mathematica* routinely carries at least 16 places of decimal accuracy. This is far more than what we will need in this text, where we shall report the results of computations to 6 places (1 part per million accuracy) except in those few cases where greater precision is required.

To this point, we have only modeled the Earth as traveling in a circular orbit at one Astronomical Unit (its average distance of 149.6 million kilometers) from the center of the sun. What is necessary next, is to use the computer to find all the parameters of the Earth's *elliptical* orbit about the sun, using only Newton's Law of Gravity, and a starting distance and velocity for the Earth.

1.1 Motion of the Planets about the Sun

Let's program the actual motion of the Earth about the sun. Astronomers tell us that the Earth is closest to the sun every January 3, at a distance of 147.1×10^6 km from the center of the sun, and the velocity of the Earth at that time is 30.288 km/s. Let us program in these two numbers as *initial conditions*, and see if we can find the complete orbit of the Earth about the sun.

From the plot, the orbit appears circular. However, if we **Table** the numbers, we see that the Earth actually slows down a small amount, and moves outward from the sun by a small amount, and halfway through the orbit, speeds up again, as it comes in closer to the sun. The orbit is an ellipse.

We have programmed in a starting velocity a little greater than that required for circular orbit. Therefore our starting point is the perihelion and the point of furthest excursion is the aphelion. What is worth noting is that we only specified the location and velocity of the Earth at ONE POINT.

And this, with Newton's Law of Gravity, is all you need to describe the motion of any planet or comet about the sun, or any satellite about a planet.

Now, the correct equations of motion for the Earth about the sun are

$$\ddot{x} = -\frac{GM}{x^2 + y^2}\, \text{Cos}\,\theta = -\frac{GM\,x}{(x^2 + y^2)^{3/2}} \qquad x_o = 147.1 \times 10^6 \text{ km} \qquad \dot{x}_o = 0$$

$$\ddot{y} = -\frac{GM}{x^2 + y^2}\, \text{Sin}\,\theta = -\frac{GM\,y}{(x^2 + y^2)^{3/2}} \qquad y_o = 0 \qquad \dot{y}_o = 30.288 \text{ km/s}$$

```
G = 3600^2 * 6.673 * 10^-20; M = 1.989 * 10^30;
sol1 = NDSolve[
  {x''[t] == G * M * (-x[t]) / (x[t]^2 + y[t]^2)^1.5,
   y''[t] == G * M * (-y[t]) / (x[t]^2 + y[t]^2)^1.5, x[0] == 147.1 * 10^6,
   y[0] == 0, x'[0] == 0, y'[0] == 30.288 * 3600}, {x, y}, {t, 0, 9000}]

InterpFunc1 = x /. sol1[[1]]; InterpFunc2 = y /. sol1[[1]];
tbl = Table[{InterpFunc1[t], InterpFunc2[t]}, {t, 0, 5000, 200}];
ListPlot[tbl, AspectRatio → Automatic, Prolog → AbsolutePointSize[3]]
```

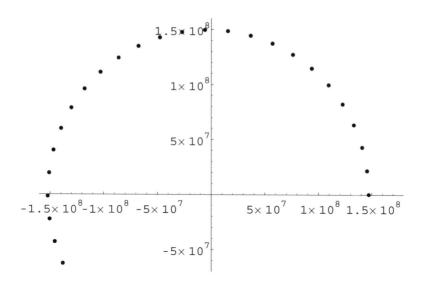

The orbit of the Earth, from January to July.

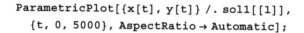

```
ParametricPlot[{x[t], y[t]} /. sol1[[1]],
   {t, 0, 5000}, AspectRatio → Automatic];
```

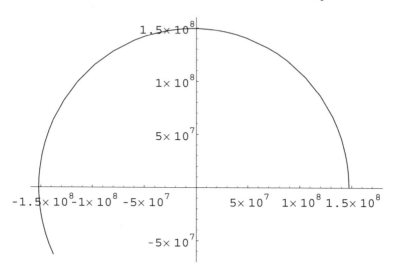

The orbit of the Earth about the sun (at 0, 0). Although the orbit appears circular, the following Table shows that it is an ellipse.

```
In[210]:= InterpFunc3 = x' /. sol1[[1]]; InterpFunc4 = y' /. sol1[[1]];
          r[t_] = √(InterpFunc1[t]^2 + InterpFunc2[t]^2)
          Vel[t_] := √(InterpFunc3[t]^2 + InterpFunc4[t]^2) / 3600
          Table[{t, InterpFunc1[t], Chop[InterpFunc2[t]],
            r[t], Vel[t]}, {t, 0, 8800, 400}] // TableForm
```

t (hr)	x (km)	y (km)	r (km)	v (km / s)
0	1.471×10^8	0	1.471×10^8	30.288
400	1.40788×10^8	4.29893×10^7	1.47205×10^8	30.2666
800	1.2242×10^8	8.22999×10^7	1.47512×10^8	30.2046
1200	9.36301×10^7	1.1461×10^8	1.47994×10^8	30.1076
1600	5.69426×10^7	1.37264×10^8	1.48607×10^8	29.9844
2000	1.55064×10^7	1.48492×10^8	1.49299×10^8	29.846
2400	-2.72117×10^7	1.47524×10^8	1.50013×10^8	29.7039
2800	-6.77342×10^7	1.34609×10^8	1.5069×10^8	29.5697
3200	-1.02851×10^8	1.10935×10^8	1.51277×10^8	29.4539
3600	-1.29847×10^8	7.84934×10^7	1.51728×10^8	29.3652
4000	-1.46679×10^8	3.99028×10^7	1.5201×10^8	29.31
4400	-1.52089×10^8	-1.79697×10^6	1.521×10^8	29.2923
4800	-1.45678×10^8	-4.3357×10^7	1.51993×10^8	29.3133
5200	-1.27919×10^8	-8.15376×10^7	1.51696×10^8	29.3715
5600	-1.00141×10^8	-1.13326×10^8	1.51232×10^8	29.4628
6000	-6.44497×10^7	-1.36152×10^8	1.50635×10^8	29.5806
6400	-2.36095×10^7	-1.48083×10^8	1.49953×10^8	29.7158
6800	1.91367×10^7	-1.48007×10^8	1.49239×10^8	29.8581
7200	6.03011×10^7	-1.35761×10^8	1.48551×10^8	29.9957
7600	9.64315×10^7	-1.12202×10^8	1.47947×10^8	30.117
8000	1.24421×10^8	-7.91801×10^7	1.47479×10^8	30.2113
8400	1.41812×10^8	-3.94194×10^7	1.47188×10^8	30.2701
8800	1.47054×10^8	3.71581×10^6	1.47101×10^8	30.2878

The above Table shows the motion of the Earth in its elliptical orbit about the sun. Notice that the maximum distance r (aphelion) is near t = 4400 hours, and the minimum distance (perihelion) at t = 0, and again at t = 8800 hours.

We may zoom in on any part of this Table to get a better read on any part of the orbit. If we wish to examine the point of furthest distance from the sun, we expand the region $(t = 4380$ to 4390 hours) to find where $y \simeq 0$.

Then if we wish to know the time of orbit completion, we expand $(t = 8760$ to 8770 hours) to find where $y \simeq 0$.

```
In[214]:= Table[{t, InterpFunc1[t], InterpFunc2[t], r[t], Vel[t]},
            {t, 4380, 4390, 1}] // TableForm
          Table[{t, InterpFunc1[t], InterpFunc2[t], r[t], Vel[t]},
            {t, 8760, 8770, 1}] // TableForm
```

t (hr)	x (km)	y (km)	r (km)	v (km / s)
4380	-1.521×10^8	312033.	1.521×10^8	29.2923
4381	-1.521×10^8	206581.	1.521×10^8	29.2923
4382	-1.521×10^8	101129.	1.521×10^8	29.2923
4383	-1.521×10^8	-4323.56	1.521×10^8	29.2923
4384	-1.521×10^8	-109776.	1.521×10^8	29.2923
4385	-1.521×10^8	-215228.	1.521×10^8	29.2923
4386	-1.521×10^8	-320680.	1.521×10^8	29.2923
4387	-1.521×10^8	-426132.	1.521×10^8	29.2923
4388	-1.52099×10^8	-531584.	1.521×10^8	29.2923
4389	-1.52099×10^8	-637035.	1.521×10^8	29.2923
4390	-1.52098×10^8	-742486.	1.521×10^8	29.2923

t (hr)	x	y	r (km)	v (km / s)
8760	1.47099×10^8	-645273.	1.471×10^8	30.288
8761	1.47099×10^8	-536237.	1.471×10^8	30.288
8762	1.47099×10^8	-427201.	1.471×10^8	30.288
8763	1.471×10^8	-318164.	1.471×10^8	30.288
8764	1.471×10^8	-209128.	1.471×10^8	30.288
8765	1.471×10^8	-100091.	1.471×10^8	30.288
8766	1.471×10^8	8946.02	1.471×10^8	30.288
8767	1.471×10^8	117983.	1.471×10^8	30.288
8768	1.471×10^8	227020.	1.471×10^8	30.288
8769	1.471×10^8	336056.	1.471×10^8	30.288
8770	1.47099×10^8	445093.	1.471×10^8	30.288

From the above two Tables, we find that the maximum excursion from the sun occurs at $t = 4383$ hours when $y \simeq 0$, $r = 152.1 \times 10^6$ km and $v = 29.292$ km/s.

Then the Earth returns to its starting position at $t = 8766$ hours when $y \simeq 0$, $x = 147.1 \times 10^6$ km, and $v = 30.288$ km/s.

The time of orbit completion is $T = 8766$ hours $= 365.25$ days, and the average distance of the Earth from the center of the sun is 149.6×10^6 km. Thus we are able to determine all the orbital parameters of the Earth's orbit. Let us now show that these numbers are correct.

From any astronomy text, we may find the following two values: the *eccentricity* of the Earth's orbit is $\epsilon = 0.01671$ and the *semi-major axis* of Earth's orbit is $a = 149.6 \times 10^6$ km

Thus the *perihelion distance* is $a(1 - \epsilon) = 147.1 \times 10^6$ km
and the *aphelion distance* is $a(1 + \epsilon) = 152.1 \times 10^6$ km

The *perihelion velocity* is found from $\frac{MG}{r1} - \frac{v^2}{2} = \frac{MG}{2a}$ $v1 = 30.288$ km/s
The *aphelion velocity* is found from $\frac{MG}{r2} - \frac{v^2}{2} = \frac{MG}{2a}$ $v2 = 29.292$ km/s

as is most easily shown using the *Mathematica* Solve command

```
In[1]:=  G = 6.673 * 10^-20; M = 1.989 * 10^30;
         a = 1.496 * 10^8; r1 = 147.1 * 10^6; r2 = 152.1 * 10^6;
         Solve[ (G M)/r1 - v^2/2 == (G M)/(2 a) , v]
         Solve[ (G M)/r2 - v^2/2 == (G M)/(2 a) , v]

Out[2]=  {{v -> -30.288}, {v -> 30.288}}

Out[3]=  {{v -> -29.2923}, {v -> 29.2923}}
```

The time (in seconds) for the Earth to complete one orbit may be found from

```
In[8]:=  T = sqrt( (4 π^2 a^3)/(G M) )

Out[8]=  3.15573 × 10^7
```

or 365.246 days. Therefore all the *Mathematica* results (of the preceding pages) are in agreement with the classical dynamics of Newton.

Let us now **Animate** the motion of the Earth about the sun.   All we need to do is **Table** the (x,y) values and display them as a data points in a **Do Loop,** incrementing every 200 hours, to show that the Earth moves in a nearly-circular orbit around the sun.

```
In[21]:= Do[ListPlot[
         Table[{InterpFunc1[t] / 10^6, InterpFunc2[t] / 10^6},
          {t, 200, i, 200}], PlotRange → {{-155, 160}, {-155, 160}},
         AspectRatio → Automatic, ImageSize → 4 * 60,
         Prolog → AbsolutePointSize[5]], {i, 200, 8800, 200}];
```

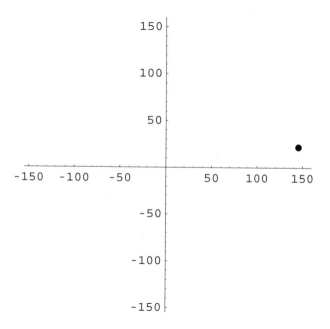

Download the file "**Motion-of-the-Planets**" to your computer and double-click on the figure to see the animation.

For more on the motion of the planets about the sun, see *Feynman's Lost Lecture.*[§] There, Richard Feynman uses arguments from plane geometry to show that the planets move in ellipses about the sun.

[§] *Feynman's Lost Lecture* edited by David and Judith Goodstein, Norton Publications, 1996

1.2 Halley's Comet near the Sun

Every 76 years[§] or so, Halley's comet makes another trip around the sun. The comet approaches to within 0.59 Astronomical Units or 88.5 million kilometers from the sun, and has a velocity of 54.45 km/s at closest approach. If we utilize this information, with Newton's Law of Gravity, we can track Halley's comet in its orbit about the sun. The physics equations are

$$\ddot{x} = - \frac{GM\, x}{(x^2 + y^2)^{3/2}} \qquad x_o = -88.5 \times 10^6 \text{ km} \qquad \dot{x}_o = 0$$

$$\ddot{y} = - \frac{GM\, y}{(x^2 + y^2)^{3/2}} \qquad y_o = 0 \qquad \dot{y}_o = -54.45 \text{ km/s}$$

Let's input these equations into *Mathematica* with time in hours and distance in kilometers.

```
M = 2 * 10^30; G = 3600^2 * 6.67 * 10^-20; (*dist in km    time in hr*)
sol = NDSolve[{x''[t] == G * M * (-x[t]) / (x[t]^2 + y[t]^2)^1.5,
    y''[t] == G * M * (-y[t]) / (x[t]^2 + y[t]^2)^1.5,
    x[0] == -88.5 * 10^6, y[0] == 0, x'[0] == 0,
    y'[0] == -54.45 * 3600}, {x, y}, {t, -9600, 9600}]

InterpFunc1 = x /. sol[[1]];   InterpFunc3 = x' /. sol[[1]];
InterpFunc2 = y /. sol[[1]];   InterpFunc4 = y' /. sol[[1]];
ListPlot[Table[{InterpFunc1[t] / 10^6, InterpFunc2[t] / 10^6},
    {t, -9600, 9600, 160}], AspectRatio → Automatic]
```

In solving the physics equations for the comet, notice that we proceed from − 9600 hours to 9600 hours, where the comet makes its closest approach to the sun at time t = 0. It does not matter to the *Mathematica* numerical solver whether we proceed backwards or forwards in time. The comet is shown on the following page from 400 days before to 400 days after its closest approach to the sun.

[§] The period of time for Halley's comet to orbit the sun varies by a few months from century to century. This was explained by Fred Whipple with his "dirty snowball theory". Each time Halley's approaches the sun, it throws off up to 25 metric tons per second of water vapor and volatile gases. This changes the orbital period in an uncertain manner. Nevertheless, Halley's comet will return sometime in 2062.

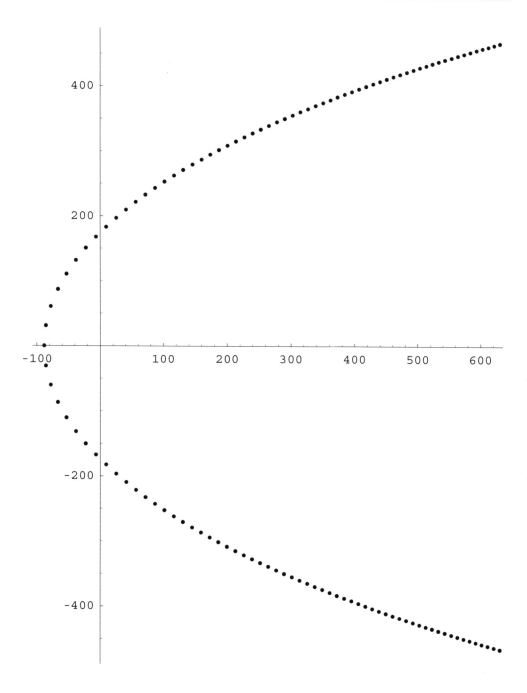

Halley's comet. From near the orbit of Jupiter, sweeping around the sun. Distances shown are in millions of kilometers. Time increment: 7 days between points.

Total time: 400 days from orbit of Jupiter to perihelion, then 400 more days to reach the orbit of Jupiter.

An interesting question that may be solved rapidly with *Mathematica* is the following: Just how long does Halley's comet spend *inside the orbit of the Earth*?   If we **Table** the above results, we can just visually scan the Table and see that the comet is within 1 A.U. of the sun for 40 days coming in, and 40 days going out, or about 80 days altogether.

```
In[7]:=  Table[{t / 24, InterpFunc1[t], InterpFunc2[t],
            √(InterpFunc1[t]^2 + InterpFunc2[t]^2) / (150 * 10^6),
            √(InterpFunc3[t]^2 + InterpFunc4[t]^2) / 3600},
          {t, -2400, 2400, 240}] // TableForm
```

t (days)	x (km)	y (km)	r (AU)	v (km / s)
-100	1.16394×10^8	2.61915×10^8	1.91075	29.6814
-90	9.42886×10^7	2.47914×10^8	1.76826	30.9192
-80	7.16813×10^7	2.32585×10^8	1.62253	32.3472
-70	4.85821×10^7	2.1564×10^8	1.47364	34.0163
-60	2.50415×10^7	1.96696×10^8	1.32189	35.9953
-50	1.19482×10^6	1.75222×10^8	1.16818	38.376
-40	-2.26446×10^7	1.50482×10^8	1.01451	41.2716
-30	-4.57804×10^7	1.21465×10^8	0.865371	44.7827
-20	-6.6705×10^7	8.69301×10^7	0.730491	48.8365
-10	-8.24177×10^7	4.59903×10^7	0.629207	52.6968
0	-8.85×10^7	0.	0.59	54.45
10	-8.24177×10^7	-4.59903×10^7	0.629207	52.6968
20	-6.6705×10^7	-8.69301×10^7	0.730491	48.8365
30	-4.57804×10^7	-1.21465×10^8	0.865371	44.7827
40	-2.26446×10^7	-1.50482×10^8	1.01451	41.2716
50	1.19482×10^6	-1.75222×10^8	1.16818	38.376
60	2.50415×10^7	-1.96696×10^8	1.32189	35.9953
70	4.85821×10^7	-2.1564×10^8	1.47364	34.0163
80	7.16813×10^7	-2.32585×10^8	1.62253	32.3472
90	9.42886×10^7	-2.47914×10^8	1.76826	30.9192
100	1.16394×10^8	-2.61915×10^8	1.91075	29.6814

Jerry Marion, in his book *Classical Dynamics* poses the same problem, and finds an analytic solution. However, solving this problem using Jerry Marion's method is a lot more work.

1.3   Voyager at Jupiter

One of the most impressive accomplishments of the United States space program is the odyssey of the Voyager satellite to the outer planets. To get there, NASA used a procedure called "gravity assist" to get the satellite up to speed. This consisted of bringing the satellite in behind Jupiter in its orbit. Then, as Jupiter moved away, the satellite got to keep most of its Kinetic Energy gain from Jupiter's gravitational acceleration.

The mass of Jupiter is 1.9×10^{27} kg, and the giant planet is moving at 13 km/s along $+ x$. At $t = 0$, Jupiter is at $(0,0)$. At any time t, Jupiter will be at $(13t, 0)$, and the satellite at (x, y). We start the satellite with an initial velocity of 10 km/s on an initial heading toward $(0, 0)$. The initial distance of the satellite will be 10^6 km from Jupiter at an angle of $40°$ below the $+x$ axis.

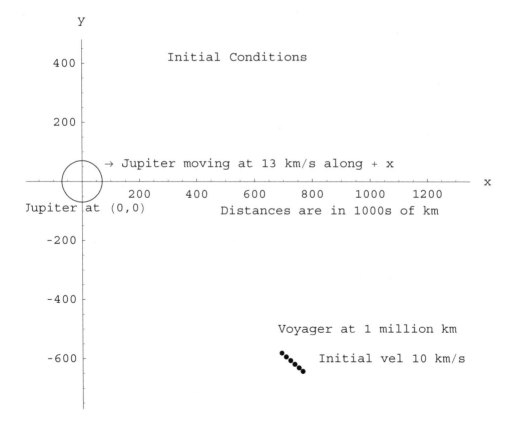

The Gravitational equations of Newton with the initial conditions are:

$$\ddot{x} = -\frac{G M (x - 13 t)}{\left((x - 13 t)^2 + y^2 \right)^{3/2}} \quad x_o = 10^6 \text{km} \times \text{Cos } 40° \quad \dot{x}_o = -10 \text{ km/s} \times \text{Cos } 40°$$

$$\ddot{y} = -\frac{G M y}{\left((x - 13 t)^2 + y^2 \right)^{3/2}} \quad y_o = -10^6 \text{ km} \times \text{Sin } 40° \quad \dot{y}_o = 10 \text{ km/s} \times \text{Sin } 40°$$

Where the mass of Jupiter and the gravitational constant are given by

$$M = M_{\text{Jupiter}} = 1.9 \times 10^{27} \text{ kg}$$

$$G = 6.67 \times 10^{-11} \frac{\text{m}^3}{\text{kg} \cdot \text{s}^2} = 60^2 \times 6.67 \times 10^{-20} \frac{\text{km}^3}{\text{kg} \cdot \text{min}^2}$$

We will program these equations into *Mathematica* with distance in kilometers, and time in minutes

```
M = 1.9 * 10^27; G = 60^2 * 6.67 * 10^-20;
sol = NDSolve[{
    x''[t] == M * G * (13 * 60 t - x[t]) / ((x[t] - 13 * 60 t)^2 + y[t]^2)^1.5,
    y''[t] == M * G * (-y[t]) / ((x[t] - 13 * 60 t)^2 + y[t]^2)^1.5,
    x[0] == 10^6 * Cos[40 π / 180], y[0] == -10^6 * Sin[40 π / 180],
    x'[0] == -10 * 60 * Cos[40 π / 180],
    y'[0] == 10 * 60 * Sin[40 π / 180]}, {x, y}, {t, 0, 1200}]
```

We may inquire of *Mathematica* as to where the spacecraft is at any time by making the following identifications.

```
In[3]:= InterpFunc1 = x /. sol[[1]];
        InterpFunc2 = y /. sol[[1]];
        InterpFunc3 = x' /. sol[[1]];
        InterpFunc4 = y' /. sol[[1]];

In[9]:= ListPlot[Table[{InterpFunc1[t] / 1000, InterpFunc2[t] / 1000},
        {t, 0, 1200, 30}], PlotRange → {{0, 900}, {-650, 750}},
        Prolog → AbsolutePointSize[4.0],
        Epilog → Circle[{.78 * 750, 10}, 70], AspectRatio → Automatic]
```

The entire trajectory of Voyager is shown on the next page, with the location of Jupiter at the point of closest approach. The points representing the satellite are at equal time intervals (every 30 minutes) and show the increase in satellite velocity. All distances shown are in thousands of kilometers.

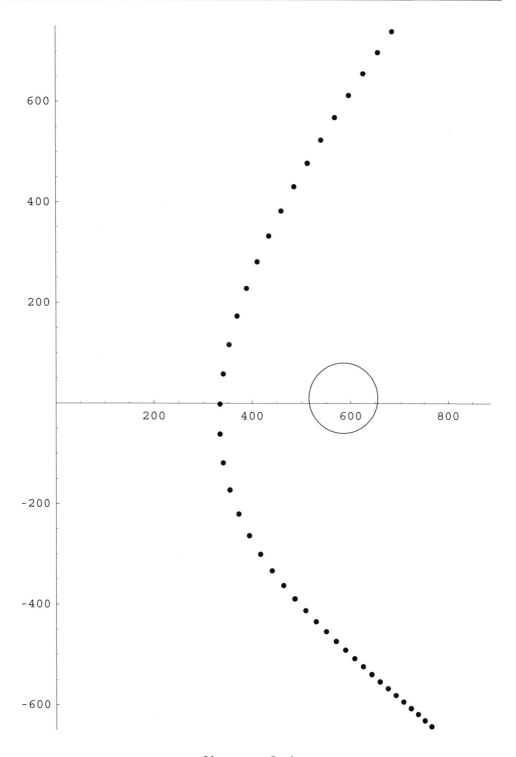

Voyager at Jupiter

Let us use the **Table** option to find the distances between the space craft and Jupiter as a function of time.

```
Table[{t, InterpFunc1[t] / 1000, InterpFunc2[t] / 1000,
    √(InterpFunc3[t]^2 + InterpFunc4[t]^2) / 60, (13.0 * 60 * t / 1000),
    √((InterpFunc1[t] - 780 t)^2 + InterpFunc2[t]^2) / 1000},
    {t, 0, 1200, 60}] // TableForm
```

The location and velocity of the satellite and its distance from Jupiter (in thousands of kilometers) are shown in the following Table. The satellite approaches to within 235,000 km of the center of Jupiter (approx 160,000 km above the cloud-tops), and picks up speed going from an initial 10 km/s to over 33 km/s on closest approach. Then, as Jupiter moves away, the satellite leaves on its new trajectory with a final velocity of approximately 28 km/s.

t (min)	x (1000 km)	y (1000 km)	v (km / s)	x (Jup) (1000 km)	d Jup (1000 km)
0	766.044	−642.788	10.	0	1000.
60	737.812	−619.086	10.4916	46.8	927.774
120	708.155	−594.104	11.0661	93.6	854.774
180	676.879	−567.552	11.7468	140.4	780.977
240	643.75	−539.03	12.5666	187.2	706.393
300	608.49	−507.975	13.5732	234.	631.095
360	570.781	−473.567	14.8379	280.8	555.297
420	530.298	−434.558	16.4692	327.6	479.507
480	486.831	−388.975	18.6327	374.4	404.898
540	440.715	−333.606	21.5597	421.2	334.177
600	394.137	−263.412	25.4325	468.	273.572
660	354.183	−172.54	29.7759	514.8	235.729
720	333.891	−61.4224	32.7412	561.6	235.847
780	340.662	57.6186	33.2933	608.4	273.868
840	368.77	172.88	32.5784	655.2	334.559
900	409.753	280.841	31.6	702.	405.315
960	458.147	381.915	30.6903	748.8	479.936
1020	510.965	477.302	29.9124	795.6	555.728
1080	566.562	568.1	29.2585	842.4	631.525
1140	623.987	655.176	28.7071	889.2	706.819
1200	682.662	739.192	28.238	936.	781.399

To better represent the entire Gravity Assist, an animated series of plots is given by the following **Do Loop**. Three stills from this Animation are shown. To see the entire Animation, you will need to access the program *VoyagerAtJupiter.nb* on the CD, and run it in *Mathematica*.

```
Needs["Graphics`MultipleListPlot`"]
Do[MultipleListPlot[
    Table[{InterpFunc1[t] / 1000, InterpFunc2[t] / 1000}, {t, 0, i, 30}],
    {{0 + .78 * t, 10}} /. {t → i},
    PlotRange → {{0, 1000}, {-680, 800}}, ImageSize → 6 * 72,
    SymbolShape → {PlotSymbol[Star], MakeSymbol[Circle[{0, 0}, 70]]},
    AspectRatio → Automatic], {i, 0, 1200, 30}]
```

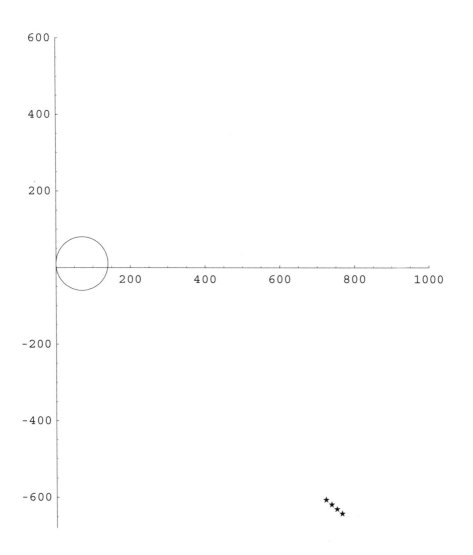

Voyager at the beginning of its approach to Jupiter.

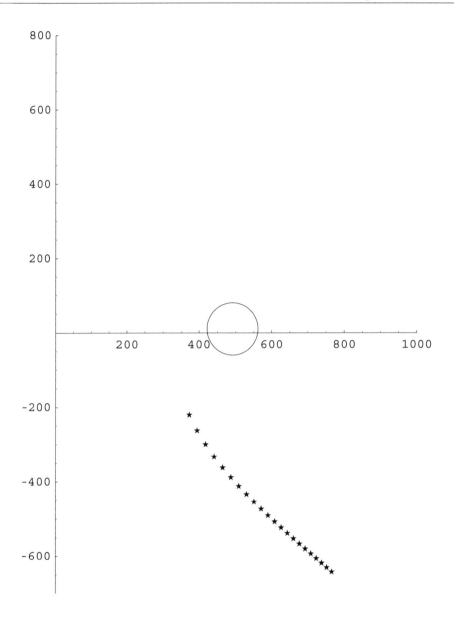

The Voyager spacecraft speeds up as it approaches Jupiter, which moves away at 13 km/s, thus allowing Voyager to keep most of its velocity gain.

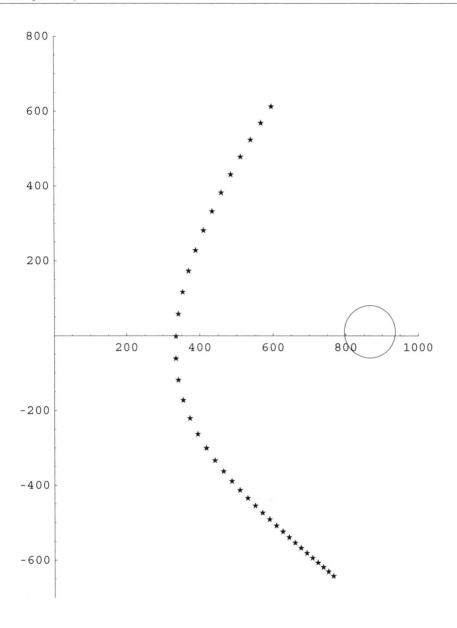

For more on the Gravitational Assist (or Slingshot Effect) see the article "Gravitational Assist in Celestial Mechanics" by James van Allen in the May 2003 *American Journal of Physics,* pages 448-451.

1.4   Baseball with Air Friction

The flight of a baseball once struck by the bat, is determined by the velocity of the ball, gravity, and air-frictional drag. If a bat speed of approximately 90 miles/hour can be achieved (by a very strong major-league batsman), then an analysis of momentum exchange would suggest an initial velocity of approximately 220 ft/s (150 mph) for the baseball. If we assume an initial angle of 30° for the baseball off the bat, then with an assumed constant drag coefficient $C_D = 0.4$, air density $\rho = 0.077$ lb/ft^3, diameter of baseball $= 0.238$ ft, and weight of the baseball $mg = 0.3125$ lb. The drag force is $D = \frac{1}{2} \rho A C_D v^2$

In units of feet and seconds, the Equations of Motion of the baseball are

$$mx'' = -D * \text{Cos}\,\Theta = -\frac{1}{2}\rho A C_D v^2 \frac{v_x}{v} = -\frac{1}{2}\rho A C_D v\, v_x$$

$$my'' = -mg - D * \text{Sin}\,\Theta = -mg - \frac{1}{2}\rho A C_D v^2 \frac{v_y}{v} = -mg - \frac{1}{2}\rho A C_D v\, v_y$$

or, in most compact *Mathematica* form

$$x'' = -.00219\, x'\sqrt{x'^2 + y'^2} \qquad x_o = 0 \qquad x'_o = 220 * \text{Cos}\,30°$$

$$y'' = -32 - .00219\, y'\sqrt{x'^2 + y'^2} \qquad y_o = 0 \qquad y'_o = 220 * \text{Sin}\,30°$$

Mathematica easily solves these non-linear equations

```
sol = NDSolve[{y''[t] == -32 - .00219 y'[t] * √x'[t]^2 + y'[t]^2 ,
    x''[t] == -.00219 x'[t] * √x'[t]^2 + y'[t]^2 , x[0] == y[0] == 0,
    x'[0] == 190, y'[0] == 110}, {x, y}, {t, 0, 5.0}]
InterpFunc1 = x /. sol[[1]]; InterpFunc2 = y /. sol[[1]];
InterpFunc3 = x' /. sol[[1]]; InterpFunc4 = y' /. sol[[1]];
tbl = Table[{InterpFunc1[t], InterpFunc2[t]}, {t, 0, 5, .2}];
ListPlot[tbl, Prolog → AbsolutePointSize[4], AspectRatio → 0.3];
```

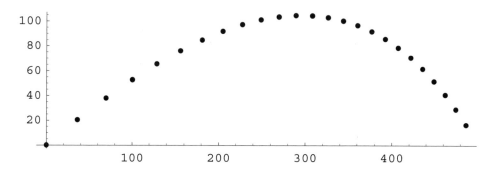

Does this baseball have home-run distance? If we **Table** the results, we will find the distance the baseball travelled x = 498 ft, and the time t = 5 seconds, for it to return to Earth (when y ≃ 0).

```
Table[{t, Chop[InterpFunc1[t]], Chop[InterpFunc2[t]],
    √(InterpFunc3[t]^2 + InterpFunc4[t]^2)}, {t, 0, 5, .2}] // TableForm
```

The (x,y) coordinates and velocity of the baseball during its 5-second flight

t (s)	x (ft)	y (ft)	v (ft / s)
0	0	0	219.545
0.2	36.2901	20.389	197.42
0.4	69.5674	37.8575	178.966
0.6	100.331	52.774	163.365
0.8	128.967	65.4222	150.043
1.	155.779	76.0249	138.582
1.2	181.01	84.7595	128.679
1.4	204.856	91.7695	120.107
1.6	227.48	97.1723	112.695
1.8	249.012	101.065	106.313
2.	269.564	103.529	100.859
2.2	289.227	104.634	96.2518
2.4	308.078	104.44	92.4257
2.6	326.18	102.999	89.3221
2.8	343.587	100.357	86.8871
3.	360.344	96.5586	85.0684
3.2	376.487	91.6422	83.8133
3.4	392.048	85.6462	83.0677
3.6	407.053	78.6075	82.7761
3.8	421.524	70.5624	82.8825
4.	435.479	61.5473	83.3315
4.2	448.933	51.5985	84.0689
4.4	461.9	40.7528	85.0437
4.6	474.394	29.0473	86.2083
4.8	486.424	16.5197	87.5196
5.	498.002	3.20752	88.9388

We can visualize the flight of the baseball a little better if we **Animate** the plot, and put a center-field wall and an outfielder in the picture. Then, with the wall at 420 ft, and the outfielder running back at 20 ft/s, we will find that the batter is indeed successful in hitting a home-run. To see the entire Animation, access *BaseballwithAirFriction* on the CD, and run it in *Mathematica*.

Notice that to plot multiple objects in the same picture that we need to call "Multiple List Plot". Then the **Do-Loop** programming proceeds as follows

Do-Loop

```
Needs["Graphics`MultipleListPlot`"]
Do[MultipleListPlot[
  Table[{InterpFunc1[t], InterpFunc2[t]}, {t, 0, i, .2}],
  {{300 + 20 * t, 1}} /. {t → i}, Prolog → Rectangle[{420, 0}, {430, 12}],
  PlotRange → {{0, 500}, {0, 210}},
  SymbolShape → {PlotSymbol[Star], PlotSymbol[Box]}], {i, 0, 5}]
```

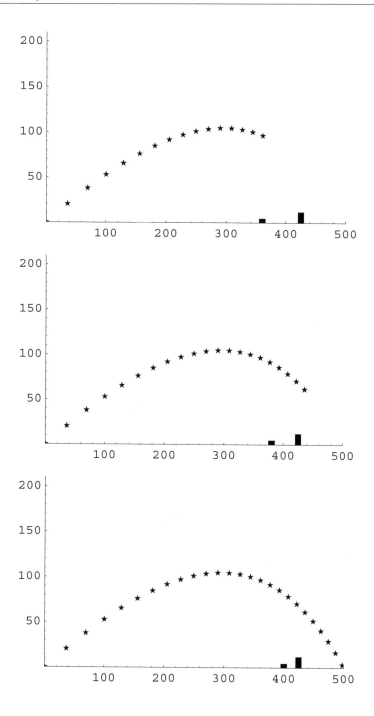

For more on batting, pitching, and fielding, see Robert Adair's 2002 book *The Physics of Baseball*, Harper-Collins, 3rd edition.

1.5 Follow the Bouncing Ball

One of the most famous pictures in Physics is the cover of the PSSC (Physical Science Study Committee) Physics text: the time-lapse photo of the bouncing ball. Let's say we drop a superball with coefficient of restitution $z = 0.9$ from a height of 8 feet. Then it is easy enough to use the energy equation $\frac{1}{2}mv^2 = mgh$ to find the velocity at impact and $h = \frac{1}{2}gt^2$ to find the time of fall.

If we let the floor supply the necessary change in velocity at each impact time, and also give the ball a horizontal velocity of 5 ft/s, so we can visualize the motion, then we have the equations of motion of the superball. (At each impact, the ball springs back with 90% of its vertical velocity. The horizontal velocity stays the same. Note the time between impacts decreases with each bounce.)

The equations of motion are as follows:

$$y' = v_y^* - gt^* \qquad y_0 = 8 \qquad \text{where time and upward velocity}$$
$$\text{are adjusted}$$
$$x' = 5 \qquad\qquad x_0 = 0 \qquad \text{to the start of each bounce}$$

```
sol = NDSolve[
    {y'[t] == 1 * Which[t < .707, 0, 0.707 ≤ t ≤ 1.98, 20.36, t > 1.98, 18.32] -
        32 * (t - Which[t < .707, 0, 0.707 ≤ t ≤ 1.98, .707, t > 1.98, 1.98]),
    x'[t] == 5, y[0] == 8, x[0] == 0}, {x, y}, {t, 0, 5}];
InterpFunc1 = x /. sol[[1]]; InterpFunc2 = y /. sol[[1]];
InterpFunc3 = y' /. sol[[1]]
ParametricPlot[{x[t], y[t]} /. sol, {t, 0, 3}, AxesLabel → {"x", "y"}];
```

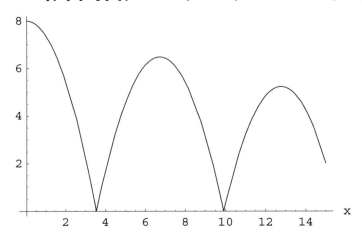

The important thing to notice in the above programming is that the time and upward velocity need to be set to appropriate values at the beginning of each bounce.This can be accomplished in *Mathematica* by using the **Which** command. Then we have a series of parabolas, each 81% as high as the last. Notice that the ball spends less time in the air on each successive bounce.

We have a geometric series which converges in time and horizontal distance travelled.

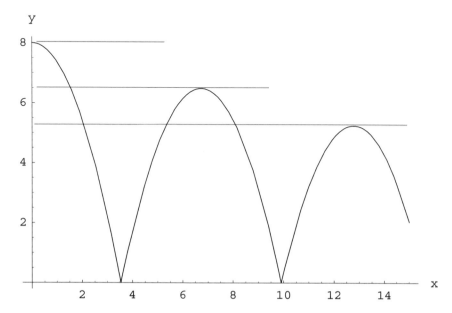

```
In[46]:= tbl = Table[{InterpFunc1[t], InterpFunc2[t]}, {t, .1, 3, .1}];
         ListPlot[tbl, Prolog → AbsolutePointSize[4], AspectRatio → 0.45]
```

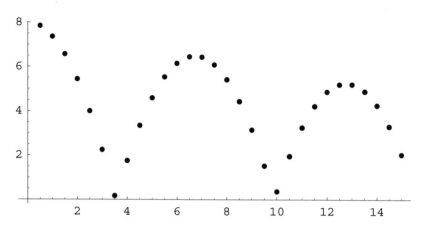

Each bounce is 81% the height of the last

```
Table[{t, InterpFunc1[t], InterpFunc2[t], InterpFunc3[t]}, {t, 0, 3, .1}]
ListPlot[tbl, Prolog → AbsolutePointSize[4], AspectRatio → 0.3]
```

t (s)	x (ft)	y (ft)	y' (ft / s)
0	0.	8.	0
0.1	0.5	7.84	-3.2
0.2	1.	7.36	-6.4
0.3	1.5	6.56	-9.6
0.4	2.	5.44	-12.8
0.5	2.5	4.	-16.
0.6	3.	2.24	-19.2
0.7	3.5	0.16	-22.4
0.8	4.	1.75	17.38
0.9	4.5	3.33	14.18
1.	5.	4.59	10.98
1.1	5.5	5.53	7.784
1.2	6.	6.15	4.584
1.3	6.5	6.44	1.384
1.4	7.	6.42	-1.81
1.5	7.5	6.08	-5.01
1.6	8.	5.42	-8.21
1.7	8.5	4.44	-11.4
1.8	9.	3.14	-14.6
1.9	9.5	1.51	-17.8
2.	10.	0.35	17.68
2.1	10.5	1.96	14.48
2.2	11.	3.24	11.28
2.3	11.5	4.21	8.08
2.4	12.	4.86	4.88
2.5	12.5	5.19	1.68
2.6	13.	5.20	-1.52
2.7	13.5	4.88	-4.72
2.8	14.	4.25	-7.92
2.9	14.5	3.30	-11.1
3.	15.	2.03	-14.3

Technically, the ball has completed 2 bounces when it hits the floor for the 3rd time.

The time to fall from $h_o = 8$ ft to the floor is $t_o = \sqrt{\dfrac{2\,h_o}{g}} = .707$ s

The time to complete the first bounce is $t_1 = 2\,(.9)\,t_o$

The total time for N bounces is $\quad T = t_o + \displaystyle\sum_{N=1}^{N} 2\,(.9^N)\,t_o$

and for 10 bounces,

$$In[113]:= \quad T = .707 + \sum_{N=1}^{10} (2 * (.9)^N * .707)$$

$Out[113]= \quad 8.99572$

Thus, the coefficient of restitution can be found for any ball from a known initial height by using a stopwatch, and finding the time to make N complete bounces.

As an example, if it takes 9 seconds for 10 complete bounces, then we use *Mathematica* to find z, the coefficient of restitution, when the ball is dropped from 8 ft.

$$In[2]:= \quad \mathbf{FindRoot}\!\left[9 == .707 + \sum_{N=1}^{10} (2 * (z)^N * .707),\ \{z,\ .8\}\right]$$

$Out[2]= \quad \{z \to 0.9001\}$

Let's also find the total time for the ball to stop bouncing, and how far it travels along x. Using *Mathematica*, we can evaluate an infinite series

$$In[15]:= \quad TF = .707 + \sum_{N=1}^{\infty} (2 * (.9)^N * .707);$$

```
        XF = 5 * TF;
        Print ["Total Time TF→ ", TF,
          "s  Distance Travelled XF→ ", XF, " ft"]
```

> Total Time TF→ 13.433s   Distance Travelled XF→ 67.165 ft

For an animation of the Bouncing Ball, select the Animation *Bouncing-Ball.nb* on the CD, and run it in *Mathematica*.

1.6 Sky-Diving with a Parachute

An 80-kg skydiver (with parachute !) jumps out of an airplane at several thousand meter altitude, with a 0.8 drag coeff and a 1 m² surface area before parachute deployment, and a 1.33 drag coefficient with a parachute area of 40 m² after deployment. Let us say that the sky-diver gets to enjoy 17 seconds of free-fall, and then pulls the ripcord. The parachute takes 5 seconds to fully deploy. How far has the skydiver fallen in 17 seconds? In 22 seconds? What is the velocity of the skydiver just before parachute deployment? And just after? And what deceleration does the skydiver experience in gees? These are all questions that can be rapidly answered with *Mathematica*.

The Equations of Motion of the sky-diver are

$$mv' = mg - \tfrac{1}{2}\rho A^* C_D^* v^2 \qquad v_o = 0 \qquad \qquad \text{*Where A and } C_D \text{ change}$$
$$mx'' = mg - \tfrac{1}{2}\rho A^* C_D^* x'^2 \qquad x_o = 0 \qquad x'_o = 0 \qquad \text{value at t = 17 s}$$

$$m = 80 \text{ kg, air density } \rho = 1.0 \text{ kg} / \text{m}^3, \ C_D = 0.8 \, , \ A = 1 \text{ m}^2, \ g = 9.8 \text{ m} / \text{s}^2$$

Let's first of all look at the approach to terminal velocity during the first 17 seconds. Since the equations are reasonably simple, we will use **DSolve** and look for an analytic solution.

```
DSolve[{80 v'[t] == 80 * 9.8 - .40 v[t] * v[t], v[0] == 0}, v[t], t] ;
Plot[Evaluate[v[t] /. %, {t, 0, 17}], AxesLabel → {"T(s)", "V(m/s)"}];
```

Out[41]= {{v[t] → 44.2719 Tanh[0. + 0.221359 t]}}

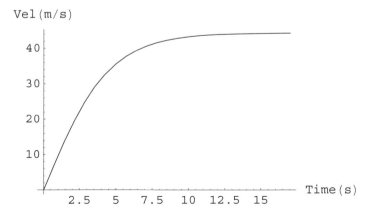

The approach to terminal velocity (44 m/s) in first 17 seconds of free-fall.

Now **DSolve** the second differential equation in x to see how far we've fallen in 17 seconds.

```
sol = DSolve[{80 x″[t] == 80 * 9.8 - .4 x′[t] * x′[t], x[0] == x′[0] == 0}, x, t]
Plot[Evaluate[x[t] /. sol, {t, 0, 17}], AxesLabel → {"Time", "Dist(m)"}]
```

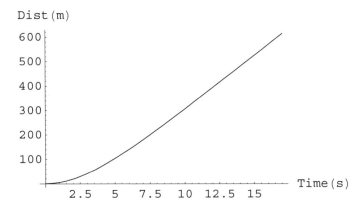

Before deploying the parachute, the skydiver has fallen a distance of six football fields in the first 17 seconds, and now has a velocity of 44 m/s (almost 100 miles per hour).

Let's **Table** the results for distance fallen, velocity, and acceleration.

```
sol = NDSolve[{80 x″[t] == 80 * 9.8 - .4 x′[t] * x′[t],
    x[0] == x′[0] == 0}, x, {t, 0, 17}];
Table[{t, x[t] /. %, x′[t] /. %, x″[t] /. %}, {t, 0, 17, 1}] // TableForm
```

t (s)	x (m)	v (m / s)	a (m / s^2)
0	0	0	9.8
1	4.8605	9.64301	9.33506
2	18.9913	18.4125	8.1049
3	41.1952	25.7251	6.49109
4	69.8903	31.3947	4.87188
5	103.477	35.5471	3.48201
6	140.572	38.4633	2.40286
7	180.095	40.4515	1.61839
8	221.256	41.7796	1.07232
9	263.504	42.6548	0.702818
10	306.465	43.2265	0.457369
11	349.89	43.5976	0.29626
12	393.617	43.8376	0.191325
13	437.537	43.9925	0.123313
14	481.583	44.0922	0.0793827
15	525.71	44.1564	0.0510641
16	569.888	44.1977	0.0328289
17	614.1	44.2242	0.021097

We are now ready to program the entire jump. Here we use the **Which** command in *Mathematica* to tell us the effective area and drag coefficient as the sky-diver proceeds through free-fall, parachute deployment, and final descent to Earth.

```
sol1 = NDSolve[
    {80 x″[t] == 80 * 9.8 - x′[t]^2 * Which[t ≤ 17, 0.40, 17 < t < 22,
        0.4 + 26.2 (t - 17) / 5, t ≥ 22, 26.6], x[0] == x′[0] == 0},
        x, {t, 0, 25}]; InterpFunc4 = x /. sol1[[1]];
Plot[x[t] /. sol1, {t, 0, 25}, AxesLabel → {"Time(s)", "x(m)"}];
Plot[x'[t] /. sol1, {t, 0, 25}, AxesLabel → {"T(s)", "Vel(m/s)"}];
Plot[(x''[t] / 9.8) /. sol1, {t, 0, 25}, AxesLabel → {"Time", "Acc"}]
```

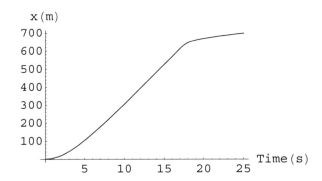

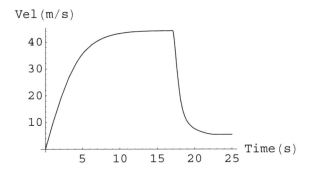

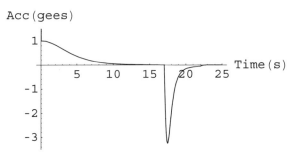

```
sol2 = NDSolve[
   {80 v'[t] == 80 * 9.8 - v[t]^2 * Which[t < 17, 0.4,
       17 ≤ t ≤ 22, 0.4 + 26.2 * (t - 17) / 5, t > 22, 26.6],
    v[0] == 0}, v, {t, 0, 25}];
Acc[t_] = 9.8 - (v[t] /. sol2)^2 * Which[t < 17, 0.4,
       17 ≤ t ≤ 22, 0.4 + 26.2 (t - 17) / 5, t > 22, 26.6] / 80;
Table[{t, x[t] /. sol1, v[t] /. sol2, Acc[t]}, {t, 17, 23, .2}] //
   TableForm
```

t (s)	x (m)	v (m / s)	a (m / s^2)
17	614.1	44.2242	0.0210971
17.2	622.784	41.8734	−21.9362
17.4	630.635	36.3202	−31.3578
17.6	637.267	30.0442	−30.1877
17.8	642.705	24.5242	−24.7226
18.	647.155	20.1737	−18.892
18.2	650.847	16.8977	−14.0706
18.4	653.971	14.4657	−10.4352
18.6	656.674	12.6575	−7.79128
18.8	659.064	11.2997	−5.89217
19.	661.216	10.2649	−4.53017
19.2	663.185	9.46227	−3.54959
19.4	665.012	8.8272	−2.83854
19.6	666.724	8.31421	−2.31781
19.8	668.343	7.89114	−1.93168
20.	669.885	7.5352	−1.64102
20.2	671.361	7.23021	−1.41842
20.4	672.78	6.96459	−1.2447
20.6	674.149	6.72998	−1.10645
20.8	675.473	6.52027	−0.994297
21.	676.758	6.33096	−0.901645
21.2	678.007	6.15863	−0.823833
21.4	679.223	6.00066	−0.757531
21.6	680.408	5.85501	−0.700322
21.8	681.565	5.72005	−0.650428
22.	682.697	5.59444	−0.606516
22.2	683.806	5.50872	−0.290053
22.4	684.903	5.46756	−0.139836
22.6	685.994	5.44768	−0.067678
22.8	687.083	5.43805	−0.0328166
23.	688.17	5.43338	−0.015927

A final review of the data shows that the sky-diver was slowed from an initial downward velocity of 44.2 m/s (99 mph) to 5.4 m/s (12 mph) by deploying the parachute. The maximum deceleration experienced by the skydiver during parachute deployment was about 3.2 g.

1.7 Challenge Problems

1. Changing Satellite Speeds

   Show that if a satellite is moving at 20 km/s on a straight-line approach to pass in front of Jupiter from 10^6 kilometers out, then after emerging from Jupiter's gravity, the satellite will be slowed. [Start the satellite from the same location as in Section 1.3, but change the satellite's initial velocity to $v_y = 20$ km/s, $v_x = 0$.]

2. Drag Racing

   A drag racer accelerates by pushing the huge rear tires against the pavement. Mike Dunn's 1999 record for time and velocity is $t = 4.5$ s, $v = 320$ mph $= 143$ m/s, in traveling the quarter mile ($x = 400$ m). Assuming the dragster maintains the same traction all the way, and air–and–road friction increases as v^2, find the coefficient of friction μ and the terminal velocity v_t that duplicates Mike Dunn's record run, using the equations

   $$v' = \mu g - kv^2 \quad \text{and} \quad x'' = \mu g - kx'^2$$

3. Solar Sailing

   A novel idea for moving freight from Earth to Mars is the solar sail. The outward velocity of the solar-ship is

   $$u = \frac{dr}{dt} = \frac{(2\, r^{-1/2}\, r_o\, (\gamma \cdot \mathrm{Sin}\, x \cdot \mathrm{Cos}^2 x))}{(a - \gamma \cdot \mathrm{Cos}^3 x)^{1/2}}$$

   For a 1000-kg ship with sail area 20,000 m^2 starting at $r = r_o = 1$ AU $= 1.5 \times 10^{11}$ m from the sun, the acceleration of the ship is $\gamma = 0.00018$ m/s^2, and $a =$ the Sun's gravity $= 0.00592$ m/s^2.

   Find the best angle x for the sail (maximize u). Then find the time required to travel by solar sail from the orbit of Earth (1 AU) to the orbit of Mars (1.5 AU).

4. Lunar Lander

"Lunar Lander" is a popular program for the small computer. The program simulates landing on the moon in a constant gravity field. You are given an initial altitude h , downward velocity v_o and a fuel supply S. By a series of rocket fuel burns you must try to land on the moon's surface at zero velocity before you run out of fuel.

See if you can devise a way to get your 2000 kg craft down safely from an initial height of 10 km if your initial velocity is 100 m/s downwards and you have 200 kg of fuel. Burn rate is proportional to thrust, and burning 1 kg/s of fuel results in an upward thrust of 4000 N.

Your maximum fuel burn rate is 2 kg/s. Assume for simplicity the mass of the landing craft is constant.   The moon's gravity near the surface is $g_{moon} = 1.7 \ m/s^2$.

5. Mickey Mantle Home Run

In 1963, Mickey Mantle hit a home run into the third deck at Yankee Stadium. The baseball struck the facade over the third deck some 106 feet above the playing field, and 365 feet from home plate. If the (x, y) coordinates of the ball were (365, 107) then this would have been the first home run ever hit out of Yankee Stadium. How far would this baseball have travelled if it had just cleared the roof of the stadium?   [To match the location of Mantle's home run, adjust the initial angle in Section 1.4.]

6. Satellite Orbits

A satellite with a rocket motor is in a 200-km high orbit about the Earth. If the rocket motor is fired briefly, giving the satellite an 8% increase in forward velocity, what is the new apogee of the orbit?   The mass of the Earth is $M_{Earth} = 6 \times 10^{24}$ kg.

7. Jupiter's Moon

Ganymede is 10^6 km from Jupiter. Assume a circular orbit. Write the equations of motion of Ganymede about Jupiter. Find the orbital period and Animate the plot. The mass of Jupiter is $M_{Jupiter} = 1.9 \times 10^{27}$ kg. [Just to make things interesting, put in Jupiter's motion at 13 km/s.]

8. Golf

The flight of a golf ball once struck off the tee, is determined by the velocity of the ball, gravity, lift, and air-frictional drag. If a driver speed of approximately 120 miles/hour can be achieved, then an analysis of momentum exchange would suggest an initial velocity of approximately 220 ft/s (67 m/s) for the golf ball. If we assume an initial angle of 10° for the golf-ball off the tee, then with an assumed constant drag coefficient $C_D = 0.32$, air density $\rho = 0.077$ lb/ft^3, diameter of golfball = 0.140 ft, and weight of the golfball $mg = 0.10125$ lb.

During time-of-flight, the drag force is $D = 0.5\rho A C_D v^2$. The Magnus (Lift) force is $L = k(\omega \times v)$ and at $\omega = 3000$ rpm, $L = .000741\ v$.

In units of feet and seconds, the Equations of Motion of the golfball are

$$mx'' = -D * \cos\theta - L * \sin\theta = -\frac{1}{2}\rho A C_D\ v\ v_x - .000741\ v\ \frac{v_y}{v}$$

$$my'' = -mg - D * \sin\theta + L * \cos\theta = -mg - \frac{1}{2}\rho A C_D\ v\ v_y + .000741\ v\ \frac{v_x}{v}$$

or, in most compact *Mathematica* form

$$x'' = -.00188\ x'\ \sqrt{x'^2 + y'^2} - .235\ y' \qquad\qquad x_o = 0 \qquad x'_o = 220 * \cos 10$$

$$y'' = -32.16 - .00188\ y'\ \sqrt{x'^2 + y'^2} + .235\ x' \qquad y_o = 0 \qquad y'_o = 220 * \sin 10$$

(A) Use the above equations of motion to find the in-air distance of a golfball driven with initial velocity $v_o = 200$, 220, and 240 ft/s, at an initial angle of 10° off the tee.

(B) Then find the approximate distance of a Tiger Woods drive, with initial velocity $v_o = 260$ ft/s, at an initial angle of 10° off the tee.

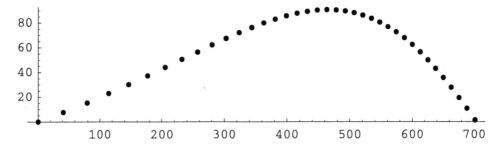

The flight of the golf ball with lift and drag,

initial vel 220 ft / s, spin 3000 rpm, initial angle 10 °

Chapter 2 Vibrations and Waves

2.1 Damped and Driven Oscillations

Part 1: Natural Response

Let's first of all take an oscillating system $m x'' + c x' + k x = 0$ consisting of a mass on a spring with a damper. We can set this system into motion either by drawing back the mass and releasing, or by starting the mass from $x = 0$ with a velocity, say $x' = 1$ m/s.

For the system $m = 1$ kg, $c = 1$ N·s/m, $k = 16$ N/m, the mass oscillates, but *not* with an angular frequency of 4 rad/s. This is shown by using **DSolve** in *Mathematica*.

```
sol = DSolve[{x''[t] + x'[t] + 16 x[t] == 0, x[0] == 0, x'[0] == 1.0},
    x[t], t] // Chop
Plot[Evaluate [x[t] /. sol, {t, 0, 6}]];
```

Out[9]= $\left\{\left\{x[t] \to 0.251976\, e^{-t/2} \sin\left[\frac{3\sqrt{7}\,t}{2}\right]\right\}\right\}$

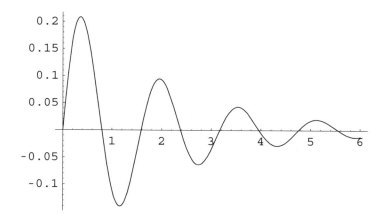

Notice that the angular frequency of vibration is $\omega = \frac{3\sqrt{7}}{2} = 3.968$ rad/s. The mass goes from zero to first maximum in a time of 0.364 seconds.

```
x[t_] = .251976 * Exp[-t / 2] * Sin[1.5 √7 t];
Maximize[{x[t], 0 < t < 1}, t]
```

Out[12]= $\{0.208377, \{t \to 0.364224\}\}$

The height at first maximum is 0.2083 meters, as found from

In[13]:= **x[t_] = .251976 Exp[-t / 2] Sin[$\frac{3\sqrt{7}}{2}$ t];**

   x[.364224]

Out[14]= 0.208377

Let's plot the above equation with an exponential envelope

In[40]:= **Plot[{x[t] /. sol, .251976 Exp[-t / 2],**
      -.251976 Exp[-t / 2]}, {t, 0, 6},
      PlotStyle → {GrayLevel[0], GrayLevel[.5], GrayLevel[.5]}];

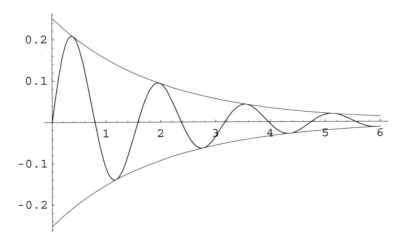

Exponential decay of an underdamped system.

An interesting question is how much does the amplitude decrease *per cycle*? All we have to do is set $\frac{3\sqrt{7}}{2}$ t $= 2\pi$. This gives the time for the mechanical system to go through one cycle. And t = 1.5832 s. Therefore the motion decays by a factor of $e^{-t/2} = .453$ from one peak to the next.

If we write this as $e^{-\delta}$ then δ is called the *logarithmic decrement*.

We now know everything about the natural response of an underdamped spring-mass-damper system. The same analysis may be applied to an overdamped or a critically damped system.

We will now use *Mathematica* to examine the steady-state response of a mechanical system to a sinusoidal driving force.

Part 2: Driven Oscillations

For a damped, driven oscillator system,

$$m\ddot{x} + c\dot{x} + kx = F\,\text{Sin}\,\omega t$$

We will see how easily *Mathematica* handles Complex Algebra.

Set the problem up so that $\gamma = \dfrac{c}{m}$ and $\omega_o = \sqrt{\dfrac{k}{m}}$

Then, using Complex Algebra, write the driving force as $F\,e^{i\omega t}$

$$m\ddot{x} + m\gamma\dot{x} + m\,\omega_o{}^2 x = F\,e^{i\omega t}$$

We look for solutions of the form $x = A\,e^{i\omega t}$ where A is complex

Using *Mathematica*,

```
In[12]:=  Clear[x, A, t]
          x[t_] = A * Exp[I w t];
          sol =
            Solve[m * x''[t] + m g x'[t] + m w0^2 x[t] == F Exp[I w t], {A}]
          A = A /. sol[[1]];
```

$$Out[14] = \left\{\left\{A \to -\frac{F}{m\,(-i\,g\,w + w^2 - w0^2)}\right\}\right\}$$

If we multiply this result times its complex conjugate, we arrive at the steady-state amplitude response to the forcing function $F\,\text{Sin}\,\omega t$

$$A = \frac{F/m}{\sqrt{\left(\omega^2 - \omega_o{}^2\right)^2 + \gamma^2\,\omega^2}}\,\text{Sin}\,(\omega t - \psi) \quad \text{where } \psi = \text{Tan}^{-1}\left(\frac{\gamma\,\omega}{\omega^2 - \omega_o{}^2}\right)$$

The amplitude of vibration A for the system $m = 1$, $\gamma = 0.5$, $\omega_o = 10$, $F = 1$ is shown on the next page, as is the phase lag, ψ.

The amplitude at resonance $\omega = \omega_o = 10$ is

$$A = \frac{1/1}{\sqrt{0^2 + .5^2 \cdot 10^2}} = 0.2 \text{ and the phase lag is } \psi = \text{Tan}^{-1}\left(\frac{\gamma\,\omega}{0}\right) = 90°.$$

In[16]:= **rule = {m → 1, F → 1, w0 → 10, g → 0.5}**
        Plot[Abs[A /. rule], {w, 0, 15}]

Out[16]= {m → 1, F → 1, w0 → 10, g → 0.5}

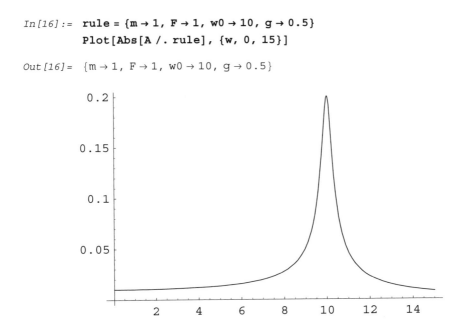

Amplitude response vs driving frequency ω

The Q is the sharpness of the resonance peak. It is numerically equal to $\frac{\omega_0}{\gamma}$
The Q for this curve is 20.

```
rule = {m → 1, F → 1, w0 → 10., g → 0.5};
Plot[-180 / π * ArcTan[Re[A] /. rule, Im[A] /. rule],
  {w, 0, 15}, PlotRange → {0, 180}]
Table[{w, Im[A] /. rule, Re[A] /. rule, Abs[A] /. rule,
    -180 / π * ArcTan[Re[A] /. rule, Im[A] /. rule]}, {w, 0, 15, 1}];
```

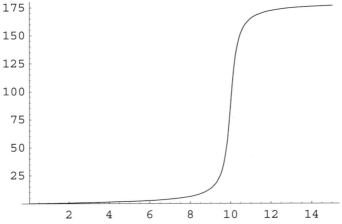

Phase lag in degrees vs driving frequency of force

If we want the *Energy* contained in the vibrating system, we need to look at A^2. Then, we find the *bandwidth* of the system, that is, the width of the curve at half-maximum

In[10]:= **F = 1; m = 1; g = 0.5; w0 = 10;**

$$\textbf{Plot}\left[\frac{(F\,/\,m)^2}{(w^2 - w0^2)^2 + g^2 * w^2}\,,\; \{w,\, 0,\, 15\}\right]$$

P lot of A^2 versus ω . The bandwidth is the width at half-maximum.

To find the *bandwidth,* find the ω values at half maximum

In[1]:= **F = 1; m = 1; g = 0.5; w0 = 10;**

$$\textbf{NSolve}\left[0.02 == \frac{(F\,/\,m)^2}{(w^2 - w0^2)^2 + g^2 * w^2}\,,\; w,\, 1\right]$$

Out[2]= {{w → 10.2409}, {w → -10.2409}, {w → 9.7403}, {w → -9.7403}}

Bandwidth is $\omega1 - \omega2$ and note that the Quality factor is $Q = \frac{\omega_o}{\omega_1 - \omega_2}$

In[1]:= $\textbf{Q} = \dfrac{10}{(10.24 - 9.74)}$

Out[1]= 20.

It is worth noting that everything that has been derived for mechanical systems is also good for electrical systems, where Force → Voltage, mass → Inductance, damping → Resistance, and the spring constant → 1/Capacitance.

As an example, consider the tuning circuit in a radio.

For AM 1000 kHz, the *bandwidth* is approx 20 kHz, and therefore its Q ≃ 50.

The value of Complex Analysis is that given a consistent driving force such as F Sin ω t, one may find the long-term or steady-state behavior of a mechanical or electrical system. However if we wish to determine <u>both</u> the initial transient and the steady-state response, then we shall have to model the system with Differential Equations. For the remainder of Chapter 2 and Chapter 3, we shall utilize only Differential Equations to model dynamic systems. We shall return to Complex Analysis in Chapter 4 to see how this very useful tool may be used in Quantum Theory.

2.2 Shock Absorbers

One of the most enjoyable things about physics is to take a physical system out in the country for a test drive. Let's say we have an old Volkswagen with worn shocks and we test-drive our vehicle on an undulating road. Let us say that the road surface rises 0.1 m and then falls below level 0.1 meters in a run of 24 meters. Then the <u>road</u> acts as a driving force on the spring-damper system, and the up-and-down contour of the road is transmitted to the mass of the chassis and the passengers. First, let's look at the road surface as it appears at v = 24 m/s (50 mph).

```
In[6]:= Clear[y, t]; L = 24; A = .1 ; v = 24; ω = 2 π * v / L;
        road = Plot[.1 * Sin[ω * t], {t, 0, 2}, AspectRatio → 0.15,
        Prolog → {Text["Time(s)", {1.88, .03}]}]
```

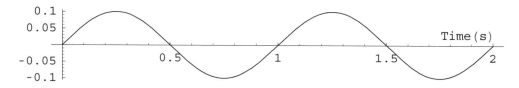

Road surface

On this "washboard" road, the road moves upwards 4 inches and then falls 4 inches below level. The road moves the damper and spring up and down, and this is transmitted to the mass of the chassis. The sum of the forces acting on the damper and spring is F(t) = k A Sin [ω t] + ω c A Cos[ω t].

The differential equation of this "driven" shock-absorber system is

$$m\,y'' + c\,y' + k\,y = k\,A\,\text{Sin}\,[\omega\,t] + \omega\,c\,A\,\text{Cos}[\omega\,t]$$

We will choose values appropriate for a car with worn shocks

$$m = 900\ \text{kg},\ \ c = 3394\ \text{N·s/m},\ \ k = 80000\ \text{N/m}$$

and now we will ask *Mathematica* to solve the equation of the road-driven system

```
m = 900; c1 = 3394; k = 80000.; A = .1;
v = 24; L = 24.; ω = 2 π * v / L; ω0 = √(k / m) ;
 sol1 = NDSolve[
    {900 y''[t] + 3394 y'[t] + 80000 y[t] == 8000 Sin[ω * t] + c1 * A * ω * Cos[ω * t],
    y[0] == 0, y'[0] == 0}, y, {t, 0, 2.0}];
InterpFunc1 = y /. sol1[[1]]; InterpFunc2 = y' /. sol1[[1]];
InterpFunc3 = y'' /. sol1[[1]];
 plot2 = Plot[Evaluate[y[t] /. sol1],
    {t, 0, 2}, PlotStyle → {{RGBColor[1, 0, 0]}},
    AspectRatio → 0.2, AxesLabel → {"time(s)", "y(m)"}];
Show [{road, plot2}, Prolog → {{Text["Road", {0.09, .13}]},
    {Text["▷", {1.3, .178}]}, {Text["Response", {.45, .17}]}},
    {Text["Time(s)", {1.75, .05}]}}}]
Table[{t, Chop[InterpFunc1[t]], Chop[InterpFunc2[t]], InterpFunc3[t],
    Chop[.1 * Sin[ω t]], Chop[InterpFunc1[t] - .1 Sin[ω t]]},
    {t, 0, 1.6, .04}] // TableForm
```

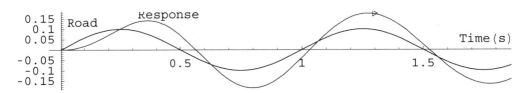

The response of the suspension-system as the automobile
travels the road.

The road "drives" the shock-absorber system, carrying the chassis "over" the top, then allowing almost "free-fall" into the canyon. What would be the effect of "taking" this series of washboard hills at 24 m/s? If we examine the Table on the next page, we will see an initial acceleration up the first hill of 2 to 4 m/s² followed by an acceleration downward at − 4 to − 7 m/s² and then up-and-down after that as the hills "drive" the suspension system up-and-down. The weakness of the shocks of our chosen automobile allow the still-good springs to "propel" the chassis and the passengers out and over the hill thus allowing a partial "free-fall" of at least the front part of the automobile and the passengers as they descend the hill.

Clearly, this all depends on the shape of the hill, the speed of the automobile, and the strength or weakness of the springs and shocks. Notice also that for this particular automobile, at this particular speed on this particular hill, that there is no danger of "bottoming" the shocks. Most suspension systems have at least 20 cm of clearance upwards and downwards before you experience that dreadful "clunk". The maximum downward deviation **z** of about 10 cm, occurs at $t = 0.84$ s, on the ascent of the second hill.

Note also that on the particular values of roadway-auto chosen above that we are not even close to resonance. For our choice of m, c, and k, the resonant frequency is $\omega_o = \sqrt{k/m} = 9.43$ whereas the driving frequency of the road is $\omega = 2\pi v/L = 6.28$ rad/s. If one were to drive the road at resonance, either $v \to 36$ m/s (75 mph !) on a $L = 24$ m road, or with $v = 24$ m/s, $L \to 16$ m (a gravel road with ruts). Running either of these roads at resonance would lead to a dreadful driving experience, however, by modeling your automobile on computer you can avoid a speeding ticket and/or bottoming-out your shocks.

t(s)	y(m)	Y'(m/s)	acc(m/s^2)	road	z = y-road
0	0	0	2.369	0	0
0.04	0.0023	0.1267	3.819	0.0248	-0.0225
0.08	0.0106	0.2926	4.307	0.0481	-0.0375
0.12	0.0257	0.4579	3.796	0.0684	-0.0427
0.16	0.0467	0.5846	2.411	0.0844	-0.0376
0.2	0.0715	0.6424	0.399	0.0951	-0.0235
0.24	0.0970	0.6126	-1.91	0.0998	-0.0027
0.28	0.1193	0.4901	**-4.17**	0.0982	0.02113
0.32	0.1350	0.2840	**-6.04**	0.0904	0.04461
0.36	0.1412	0.0151	**-7.27**	0.0770	0.06419
0.4	0.1358	-0.286	**-7.68**	0.0587	0.07708
0.44	0.1183	-0.588	**-7.22**	0.0368	0.08149
0.48	0.0892	-0.854	**-5.95**	0.0125	0.07675
0.52	0.0508	-1.055	**-4.00**	-0.012	0.06336
0.56	0.0060	-1.168	-1.60	-0.036	0.04284
0.6	-0.041	-1.181	0.983	-0.058	0.01747
0.64	-0.087	-1.090	3.494	-0.077	-0.0100
0.68	-0.127	-0.905	5.680	-0.090	-0.0368
0.72	-0.158	-0.643	**7.341**	-0.098	-0.0602
0.76	-0.178	-0.327	**8.340**	-0.099	-0.0782
0.8	-0.184	0.0140	**8.613**	-0.095	-0.0892
0.84	-0.176	0.3520	**8.169**	-0.084	-0.0925
0.88	-0.156	0.6589	**7.079**	-0.068	-0.0881
0.92	-0.125	0.9114	5.468	-0.048	-0.0768
0.96	-0.084	1.0916	3.494	-0.024	-0.0598
1.	-0.038	1.1885	1.334	0	-0.0387
1.04	0.0092	1.1982	-0.83	0.0248	-0.0156
1.08	0.0559	1.1237	-2.85	0.0481	0.00777
1.12	0.0981	0.9738	**-4.58**	0.0684	0.02968
1.16	0.1330	0.7622	**-5.92**	0.0844	0.04860
1.2	0.1585	0.5059	**-6.81**	0.0951	0.06341
1.24	0.1731	0.2237	**-7.21**	0.0998	0.07336
1.28	0.1763	-0.064	**-7.14**	0.0982	0.07810
1.32	0.1681	-0.341	**-6.62**	0.0904	0.07764
1.36	0.1493	-0.589	**-5.71**	0.0770	0.07233
1.4	0.1215	-0.794	**-4.49**	0.0587	0.06275
1.44	0.0865	-0.946	-3.05	0.0368	0.04970
1.48	0.0466	-1.037	-1.47	0.0125	0.03410
1.52	0.0044	-1.063	0.154	-0.012	0.01693
1.56	-0.037	-1.025	1.732	-0.036	-0.0007
1.6	-0.076	-0.926	3.181	-0.058	-0.0180

What if the "wavelength" of the road is 32 meters (100 feet) with the same + 4" and – 4" amplitude? We will expect less acceleration up-and-down, because the "driving frequency" ω is now less, and we are moving further away from resonance.

For this second example, $\omega = \frac{2\pi v}{L} = \frac{2\pi(24)}{32} = 4.71$ rad/s whereas $\omega_o = \sqrt{k/m} = 9.43$ rad/s.

```
In[98]:= Clear[y, t]; (* The road with L=32 m; A=.1 m; v=24 m/s*)
         m = 900; c1 = 3394; k = 80000; L = 32; v = 24; ω = 2 π * v / L; A = .1;
         sol1 =
         NDSolve[{900 y''[t] + 3394 y'[t] + 80000 y[t] == 8000 Sin[ω * t] +
             c1 * A * ω * Cos[ω * t], y[0] == 0, y'[0] == 0}, y, {t, 0, 4.0}]
         road = Plot[.1 * Sin[ω * t], {t, 0, 3.3}, AspectRatio → 0.15]
```

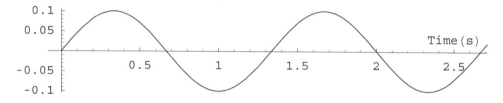

```
In[110]:= InterpFunc1 = y /. sol1[[1]]; InterpFunc2 = y' /. sol1[[1]];
          InterpFunc3 = y" /. sol1[[1]];
          plot2 = Plot[Evaluate[y[t] /. sol1], {t, 0, 3.3},
              PlotStyle → {{RGBColor[1, 0, 0]}},
              AspectRatio → 0.3, AxesLabel → {"time(s)", "y(m)"}];
          Show [{road, plot2}, Prolog → {{Text["Road", {.07, .13}]},
              {Text["Response", {.53, .13}]}},
              AxesLabel → {"time(s)", "y(m)"}]
          Table[{t, InterpFunc1[t], InterpFunc2[t], InterpFunc3[t],
              Chop[.1 * Sin[ω * t], InterpFunc1[t] - .1 Sin[ω * t]]},
              {t, 0, 2.25, .05}] // TableForm
```

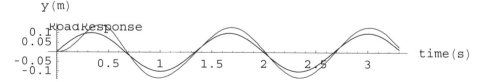

With the increased wavelength L = 32 m road, we find, in the Tabled data, less vertical movement y, and less acceleration up-and-down of the car. The "sailing" over the hill is reduced as is the "free-fall". The "sharper" the hill crest, and the greater the speed, the more pronounced is the "drop" over the hill.

t(s)	y(m)	Y'(m/s)	acc(m/s^2)	road	z = y - road
0	0	0	1.7779	0	0
0.05	0.00286	0.12533	3.0761	0.0233	-0.0204
0.1	0.01319	0.29038	3.3504	0.0453	-0.0322
0.15	0.03170	0.44374	2.6328	0.0649	-0.0332
0.2	0.05662	0.54104	1.1618	0.0809	-0.0242
0.25	0.08438	0.55346	-0.695	0.0923	-0.0080
0.3	0.11039	0.47183	-2.534	0.0987	0.01162
0.35	0.13015	0.30624	**-4.002**	0.0996	0.03046
0.4	0.14003	0.08197	**-4.852**	0.0951	0.04493
0.45	0.13794	-0.1668	**-4.982**	0.0852	0.05267
0.5	0.12353	-0.4047	**-4.426**	0.0707	0.05282
0.55	0.09817	-0.6005	-3.330	0.0522	0.04592
0.6	0.06455	-0.7326	-1.918	0.0309	0.03365
0.65	0.02615	-0.7909	-0.416	0.0078	0.01830
0.7	-0.0133	-0.7762	0.9656	-0.015	0.00232
0.75	-0.0504	-0.6989	2.0749	-0.038	-0.0121
0.8	-0.0824	-0.5747	**2.8318**	-0.058	-0.0236
0.85	-0.1074	-0.4218	**3.2262**	-0.076	**-0.0313**
0.9	-0.1244	-0.2574	**3.3034**	-0.089	**-0.0353**
0.95	-0.1332	-0.0954	**3.1425**	-0.097	**-0.0359**
1.	-0.1341	0.05436	**2.8323**	-0.1	**-0.0341**
1.05	-0.1280	0.18661	2.4514	-0.097	**-0.0308**
1.1	-0.1158	0.29926	2.0547	-0.089	-0.0267
1.15	-0.0984	0.39227	1.6682	-0.076	-0.0224
1.2	-0.0769	0.46627	1.2924	-0.058	-0.0181
1.25	-0.0521	0.52140	0.9098	-0.038	-0.0138
1.3	-0.0251	0.55675	0.4974	-0.015	-0.0094
1.35	0.00315	0.57034	0.0374	0.0078	-0.0046
1.4	0.031511	0.55961	-0.475	0.0309	0.00061
1.45	0.058677	0.52221	-1.025	0.0522	0.00642
1.5	0.083273	0.45695	-1.583	0.0707	0.01256
1.55	0.10391	0.36451	-2.104	0.0852	0.01865
1.6	0.11932	0.24798	**-2.538**	0.0951	0.02421
1.65	0.12840	0.11291	**-2.838**	0.0996	**0.02871**
1.7	0.13043	-0.0330	**-2.967**	0.0987	**0.03166**
1.75	0.12507	-0.1806	**-2.904**	0.0923	**0.03268**
1.8	0.1125	-0.3201	**-2.649**	0.0809	**0.03159**
1.85	0.0933	-0.4422	-2.205	0.0649	**0.02840**
1.9	0.0687	-0.5385	-1.623	0.0453	0.02330
1.95	0.0400	-0.6029	-0.932	0.0233	0.01668
2.	0.0090	-0.6313	-0.195	0	0.00901
2.05	-0.022	-0.6226	0.5445	-0.023	0.00085
2.1	-0.052	-0.5777	1.2394	-0.045	-0.0072
2.15	-0.079	-0.5001	1.8483	-0.064	-0.0147
2.2	-0.102	-0.3949	2.3378	-0.080	-0.0212
2.25	-0.118	-0.2687	2.6863	-0.092	-0.0264

The equations of motion allow predictability in the design of a shock-absorber system. In this section, we have examined the response of a light automobile with worn shocks on a hill.

In designing a shock absorber system for a particular automobile, it is necessary to computer-model the full range of forces and shocks which the automobile will experience. And then compare the actual response of the vehicle to on-road conditions. In automotive design, it will also be necessary to go beyond the relatively simple one-dimensional model presented here, and model all four wheels of the automobile.

2.3   Step Response

Let's say a mechanical system as shown is subject to a step-function driving force which alternates every half-second from + 100 N upwards to − 100 N downwards. What is the response of the 40 kg mass, given that the spring constant is 17720 N/m, and the damping coefficient is 88 N-s/m ? We will need to solve the differential equation

$$40y'' + 88\,y' + 17720y = F(t) \quad \text{where F(t) is the applied force}$$

$$\text{with} \quad y\,(0) = 0 \quad \text{and} \quad y'\,(0) = 0$$

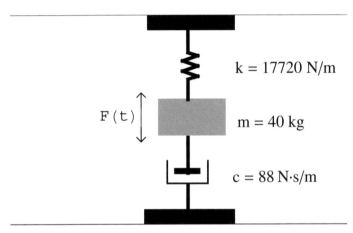

A mass-spring-damper system driven by a force F(t)

```
Force =
  Plot[200 UnitStep[Sin[2 π * t]] - 100 UnitStep[t], {t, -.2, 3.6},
    AspectRatio → 0.18, AxesLabel → {"time(s)", "Force(N)"}];
```

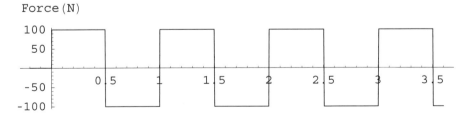

The applied force + 100 N upwards, − 100 N downward

```
In[3]:= Clear[y, t];
        sol = NDSolve[{40 y″[t] + 88 y'[t] + 17720 y[t] ==
               200 * UnitStep[Sin[2 π * t]] - 100 UnitStep[t],
              y[0] == y'[0] == 0}, y, {t, 0, 6}];
        Plot[{y[t] /. sol,
           .025 UnitStep[Sin[2 π * t]] - .0125 * UnitStep[t]}, {t, 1, 4.2},
          PlotStyle → {{RGBColor[0, 0, 0]}, {RGBColor[0, 0, 1]}},
          AspectRatio → 0.33, Prolog → {Text["λ", {2.5, .0185}],
            Text[">", {2.98, .0187}], Text["<", {2.01, .0187}]},
          Epilog → {GrayLevel[0.38], Line[{{2.6, .0188}, {2.95, .0188}}],
            Line[{{2.04, .0188}, {2.4, .0188}}]}]
```

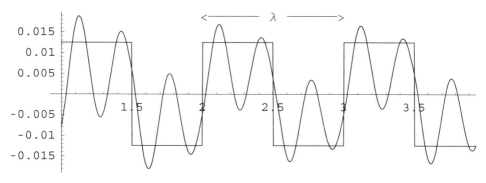

The response of a mechanical system to a square-wave input

Would you have guessed this response to a square-wave input ?

How is it possible that the response of the mechanical system can show *three* up-and-down cycles within one wavelength λ of the driving force ? The answer is very interesting. The square-wave F(t) is not a pure sine-wave or cosine-wave, but rather a composition of many sine waves.

$$F(t) = \frac{400}{\pi}\left(\ \text{Sin } 2\pi\, t + \frac{\text{Sin } 6\,\pi\, t}{3} + \frac{\text{Sin } 10\,\pi\, t}{5} + \frac{\text{Sin } 14\,\pi\, t}{7} + \ldots\ \right)$$

The very nice thing about *Mathematica* is that we can isolate the system response to just one of the driving frequencies, and just like a spectrum analyzer, see how much of the system response is due to that frequency. Since there are apparently three cycles per second in the response, let us find the system response to just $f(t) = \frac{400}{3\,\pi}$ Sin $(6\pi\, t)$.

```
In[13]:=  sol = NDSolve[{40 y''[t] + 88 y'[t] + 17720 y[t] == (400/(3 π)) * Sin[6 π * t],
              y[0] == y'[0] == 0}, y, {t, 0, 6}];
          Plot[{y[t] /. sol,
              .025 UnitStep[Sin[2 π * t]] - .0125 * UnitStep[t]}, {t, 1, 4.2},
              PlotStyle → {{RGBColor[0, 0, 0]}, {RGBColor[0, 0, 1]}},
              AspectRatio → 0.33]
```

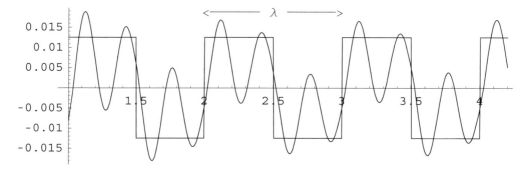

Response of the mechanical system to the complete square-wave driving force

$$F(t) = \frac{400}{\pi}\left(\mathrm{Sin}\,2\pi t + \frac{\mathrm{Sin}\,6\,\pi\,t}{3} + \frac{\mathrm{Sin}\,10\,\pi\,t}{5} + \frac{\mathrm{Sin}\,14\,\pi\,t}{7} + \ldots\right)$$

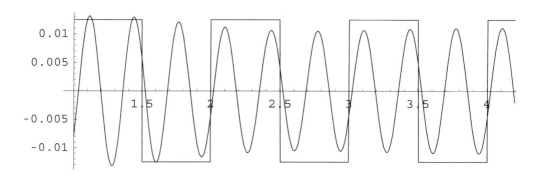

Response of the mechanical system to $f(t) = \frac{400}{3\pi}\,\mathrm{Sin}\,(6\pi\,t)$ only

Notice that in comparing this response to the full square-wave, that almost all of the response is due to the $\mathrm{Sin}(6\pi t)$ term. This is because the resonant frequency of our m = 40, c = 88, k = 17720 system is $\omega_o = \sqrt{\frac{17720}{40}} = 21.04$ rad/s whereas the driving angular frequency is $\omega = 6\pi = 18.85$ rad/s.

Thus the lion's share of the system response is to the second-harmonic 6π term, and the rest of the shape of the response is due to the other terms slightly modulating the 6π response.

2.4 Square Wave – RC Circuit

A very simple and instructive physics lab experiment is to monitor the voltage across a capacitor in an RC-circuit when a square-wave voltage is applied. When the voltage is 1 volt, the capacitor charges. When the applied voltage is 0 volts, the capacitor discharges.

For an RC circuit with $R = 5000\ \Omega$ and $C = 0.04\ \mu F$, the time constant is 0.2 milliseconds.

From Kirchoff's Law, one can write the equation for the voltage drops around the circuit. If the voltage drop across the capacitor is V, and the current in the circuit is I,

$$R\,I + V = S(t) \quad \text{where S(t) is the applied voltage}$$

$$R\frac{dQ}{dt} + V = S(t) \quad \text{and since } Q = CV,$$

$$RC\,\frac{dV}{dt} + V = S(t) \quad \text{where } S(t) = UnitStep[Sin\ \pi\ t]$$

$$V(0) = 0$$

```
In[10]:=  voltagePlot =
            Plot[UnitStep[Sin[π * t]], {t, -.8, 3.6}, AspectRatio → 0.28];
          (*Input voltage*)
```

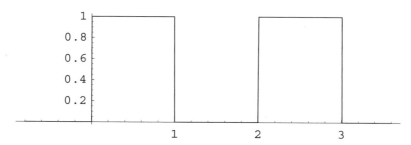

Time in milliseconds

The applied voltage: 1 volt maximum, 0 volts minimum —

```
rcsol = NDSolve[
    {.2 v'[t] + v[t] == UnitStep[Sin[π * t]], v[0] == 0}, v, {t, -1, 4}];
Plot[{UnitStep[Sin[π * t]], v[t] /. rcsol}, {t, -.8, 3.6},
  PlotRange → {0, 1}, AspectRatio → 0.30,
  AxesLabel → {"T (ms)", "Voltage"}, PlotStyle → {{Hue[2 / 3]}, {Black}}]
```

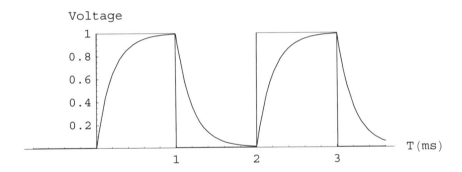

The response of an RC series circuit to a square-wave input

The response of the RC-circuit is this shark-fin charging and discharging as the voltage on the capacitor reaches an exponential charging-maximum, then falls exponentially to zero. The cycle time of S(t), the applied square-wave is 0.002 seconds, corresponding to a frequency of 500 Hz.

If one were to decrease the cycle time, by changing the frequency of the applied voltage to 2000 Hz, then the response of the circuit would be to charge-up only halfway, and give an output waveform of the following shape

```
In[48]:= Clear[v, t];
         sol =
            NDSolve[{.2 v'[t] + v[t] == UnitStep[Sin[4 π * t]], v[0] == 0},
            v, {t, -.2, 1.8}];
         Plot[{UnitStep[Sin[4 π * t]], v[t] /. sol}, {t, -.2, 1.8},
         PlotRange → {0, 1}, AspectRatio → 0.3,
         AxesLabel → {"T(ms)", "Voltage"}]
```

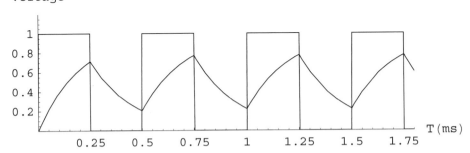

The response of the RC-circuit to a square-wave of higher frequency (2000 Hz).

The capacitor only has time to charge-up part-way before the applied voltage goes to zero.

2.5   Variable Stars

<u>The Physics of Stellar Pulsation</u>

Many giant stars pulsate. Not only does their light output vary in a periodic way, but the outer envelopes of some stars actually move in and out over millions of kilometers over a period of days. One of the most famous of these stars is Delta Cephei, a giant yellow star which oscillates with a time period of approx 5.4 days as the outer layers of the star move outward and then inward a total distance of approx 3 million kilometers. The radial velocity of the envelope exceeds 20 km/s.

It is possible to model the pulsation if we assume that some of the heat and light energy coming up from the stellar interior is absorbed by ionizing hydrogen or helium at a critical depth in the envelope. This heats and drives the envelope outwards. Then, as the hydrogen and helium cool and deionize, the stored-up energy is liberated, causing an increase in the star's luminosity. Now the de-ionized layers fall back, and the process is ready to repeat again. This one-zone model was originally proposed by Sir Arthur Eddington, who unfortunately did not have a computer or mathematical software to fully explore this concept.

We will start by assuming a certain mass of ionizable material, say $m = 10^{26}$ kg at a distance of 1.35×10^7 km from the star's center. Then the forces acting on this mass of material are gravity and pressure, where x is the distance from the stellar center

$$m\ddot{x} = -\frac{GMm}{x^2} + 4\pi\, x^2 P \qquad x_o = 1.35 \times 10^7 \text{ km} \qquad \dot{x}_o = 0$$

The pressure acting will be assumed to be *adiabatic.* That is, the initial pressure build-up drives the outer layer m outwards and the volume of the star expands from $\frac{4}{3}\pi x_o^3$ to $\frac{4}{3}\pi x^3$. And, from thermodynamics, the pressure driving the expansion falls from its original value of P_o to $P = P_o\left(\frac{x_o^3}{x^3}\right)^{5/3}$ where $P_o = 1.8 \times 10^5$ N/m^2 . (These numbers are chosen as representative for the envelope of a giant star). Combining the above equations, we have a mathematical model for a variable star of mass $M = 10^{31}$ kg

$$m\ddot{x} = -\frac{GMm}{x^2} + 4\pi\,\frac{P_o\, x_o^5}{x^3} \qquad x_o = 1.35 \times 10^7 \text{ km} \qquad \dot{x}_o = 0$$

In *Mathematica*, with distance in km and time in seconds,

```
In[1]:=  sol =
         NDSolve[{x''[t] == (-6.67 *10^11) / x[t]^2 + (1 *10^19) / x[t]^3,
             x[0] == 1.35 *10^7, x'[0] == 0}, x, {t, 0, 1000000}]
```

We will plot the motion of our variable star for 10^6 seconds, or about 12 days.

```
In[2]:=  InterpFunc1 = x /. sol[[1]]; InterpFunc2 = x' /. sol[[1]];
         Plot[Evaluate[x[t] /. sol], {t, 0, 1000000}, PlotStyle
             → {{RGBColor[1, 0, 0]}}, AspectRatio → 0.45];
```

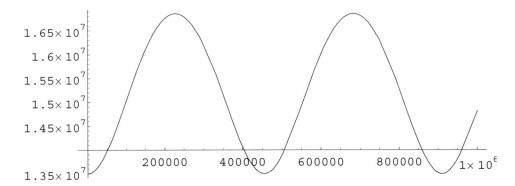

Stellar Radius as a function of time in seconds

The period of oscillation is about 450,000 seconds = 5.2 days

```
In[20]:=  Plot[Evaluate[x'[t] /. sol], {t, 0, 1000000}, AspectRatio → 0.5,
             PlotRange → {-25, 25}, PlotStyle → {{RGBColor[0, 0, 1]}}];
          tbl = Table[{t / 3600., InterpFunc1[t], Chop[InterpFunc2[t]]},
             {t, 0, 720000, 10 * 3600}] // TableForm
```

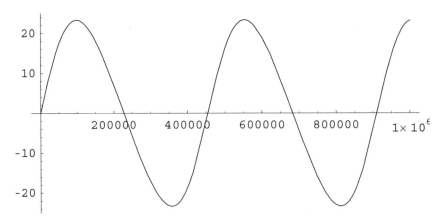

Radial velocity   outward + , inward −

Maximum velocity is approximately 23 km/s

It is reasonably clear that the simple model of Eddington produces a first approximation for stellar oscillation. The pressure that originally comes from the absorption of thermal energy in the ionization zone is converted into mechanical work in pushing the outer layer outward. Then, as the expansion proceeds, the outward pressure weakens, and the star's gravity brings the outer layers back, so the process may begin again. What is amazing, from a Physics standpoint is that the results of this reasonably simple model are fairly close to the oscillation time and expansion velocities of the star Delta Cephei.

Following is a Table of values for Stellar Radius and radial velocity. These are reasonable values and show that the one-zone model provides a good starting point for a more complicated model.

t (hrs)	R (km)	v (km/s)
0	1.35×10^7	0
10.	1.37523×10^7	13.4897
20.	1.44058×10^7	21.679
30.	1.52306×10^7	23.0846
40.	1.60035×10^7	19.1369
50.	1.65689×10^7	11.8766
60.	1.6839×10^7	2.96882
70.	1.67783×10^7	-6.30849
80.	1.63944×10^7	-14.7811
90.	1.57402×10^7	-21.0566
100.	1.49265×10^7	-23.3002
110.	1.41356×10^7	-19.5004
120.	1.36047×10^7	-8.98879
130.	1.35344×10^7	5.23639
140.	1.39542×10^7	17.2465
150.	1.46984×10^7	22.9339
160.	1.55271×10^7	22.1592
170.	1.62388×10^7	16.7849
180.	1.6704×10^7	8.75874
190.	1.68557×10^7	-0.418581
200.	1.66745×10^7	-9.54415

For the full equations of stellar pulsation, and the computer model with many zones from the stellar interior to the ionization zones, see Robert F. Christy *"Variable Stars — RealisticStellarModels"* pp 173-210 in the book <u>Stellar Evolution</u> , Hong-Yee Chiu editor, Plenum Press, 1972).

2.6 Challenge Problems

1. New York to Leningrad

A famous physics problem is to consider travel by gravity by tunneling from one city to another over great distances through the Earth. One interesting possibility would be to tunnel from New York to Leningrad, a distance of some 12,000 km over the Earth's surface or 108° of a great circle around the Earth.

Such a tunnel would reach a depth of 2640 km and would have to be impervious to the several thousand degree temperatures inside the Earth. However, if such a tunnel could be built, what would be the transit time from New York to Leningrad by a train traveling in an evacuated tunnel ? Use the very interesting fact that gravity inside the Earth is nearly constant at 10 m/s^2 to a depth of 3000 km (Adam Dziewonski and Don Anderson (1981) *Physics of Planetary Interiors* pages 297-356, "Preliminary Reference Earth Model" which gives the density, gravity, and sound speed from the Earth's surface to core).

2. Lunar Subway

Assume the moon is of constant density. Find the maximum velocity and transit time from the North pole to the South pole of the moon. The radius of the moon R_{moon} = 1700 km. The surface gravity of the moon $g_{moon} = 1.7$ m/s^2

3. Speed Bump

Let's say we're cruising along the highway in our favorite automobile when up ahead is a *speed bump*. Should we speed-up or slow down ? You had best slow-down, if you wish to preserve the integrity of your car, and the following problem will show the reason.

Assume the highway engineers have placed a 0.8-meter long, 0.1 meter high sinusoidal *speed bump* in your lane, and you approach the bump at 2 m/s, what is the suspension-system response?

Now take a computer-simulated run at the speed-bump at 4 m/s and at 8 m/s. What is the suspension-system response at each of these velocities? Assume m = 1000 kg, c = 3400 N-s/m, k = 80,000 N/m.

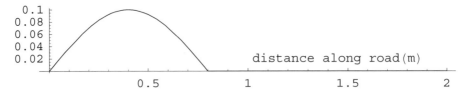

4. RLC Circuit

For an RLC-series circuit, the sum of voltage drops around the loop is equal to the driving voltage. If we have a capacitor of 2-microfarads, a resistor of 100 ohms, and an inductor of 0.5 henries, all connected in series, and an initial voltage on the capacitor of 10 volts, the differential equation of the electrical circuit with no driving voltage is

$$L \frac{dI}{dt} + R I + \frac{1}{C}Q = 0 \qquad \text{where Q is the charge on the capacitor.}$$

If we want to know the voltage on the capacitor as a function of time, let's rewrite the equation using $Q = C V$. Then,

$$LC \frac{d^2 V}{dt^2} + RC \frac{dV}{dt} + V = 0 \qquad \text{where } V_o = 10 \text{ volts and } I_o = 0.$$

Solve the above equation for V for the first 50 milliseconds after the switch is closed. What is the time-constant of this *underdamped* circuit and what is its oscillation frequency?

5. Impulse Response

Let's take a mechanical system ($m = 1$, $c = 3$, $k = 2$) at rest at $t = 0$

$$y'' + 3 y' + 2 y = F(t) \qquad y_o = 0, \quad y_o' = 0.$$

and give it a unit impulse of 1 force unit for 1 second at $t = 0$.

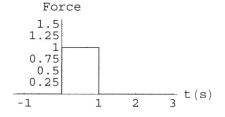

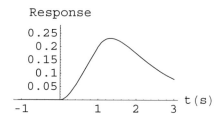

The Impulse   The Response

Now let's take the same system and subject it to a unit impulse with a force of 2 for 0.5 seconds. Then take the system and give it a unit impulse with a force of 10 for 0.1 seconds.

It's pretty clear where we're going with this. We will now give the system a unit impulse with a Dirac Delta function where $F \to \infty$ as $\Delta t \to 0$, but $F\Delta t = 1$. Do we need to modify the initial conditions to find the impulse response if the Dirac function acts at $t = 0$? What if the Dirac delta function chimes in at one-one millionth of a second after $t = 0$? Show the system response for each of these unit impulses.

Chapter 3 Building a
Differential Equation

3.1 Modeling Dynamic Systems

<u>Modeling with Differential Equations</u>

The modeling of dynamical systems with differential equations is a little bit of an art, and quite a bit of science. One endeavors to keep the equations as simple as possible while retaining the essential details of the physical system. The essentials of the modeling process are

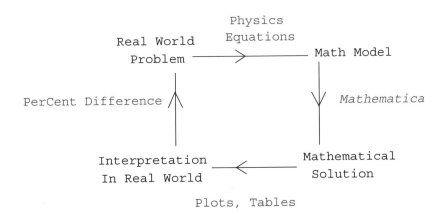

With the advent of modern high-memory personal computers and mathematical software, one is no longer restricted to linear systems. A flow-chart of the modeling process is shown above.

It is now possible to run through the modeling process, complicating the model as necessary, to obtain a reasonable match with the physical system at hand. In the following sections of this chapter, several complicated physical systems will be analyzed.

First, the Physics of running a race in Track and Field. A runner must get up to speed by accelerating, and every runner has a finite energy supply. Therefore the relevant equations that must be used are

$$F = m\,a \quad \text{and} \quad \frac{dE}{dt} = \sigma - f \cdot v$$

Second we will investigate the Physics of Binary Stars. Here it is gravity that controls the orbits of the stars, and here it is necessary to keep track of the positions and velocities of each star as they orbit one another. It is a bit of a challenge to map out the orbital dynamics of the stars, and to animate the paths of the two stars, but the Physics is that of gravitation. We start simply with two stars of equal mass, with Star 1 at (x , y) and Star 2 at $(-x , -y)$. The forces acting are then

$$Mx'' = - \frac{MGM}{(2x)^2 + (2y)^2} \frac{2x}{\sqrt{(2x)^2 + (2y)^2}} \qquad My'' = - \frac{MGM}{(2x)^2 + (2y)^2} \frac{2y}{\sqrt{(2x)^2 + (2y)^2}}$$

Third, we complicate the Binary Star problem. Now we take two stars of *unequal* mass, and after a certain time, we allow the star of greater mass to explode. In this way, we can use the computer to track the orbits of the two stars before the explosion, and then determine whether the stars remain gravitationally bound after the explosion. Again, it is gravity that controls the dynamics of the system, but the interesting aspect of the problem of the Exploding Star is that the mass of one of the components changes discontinuously at a certain time.

Fourth is an engineering system, that of Heat Exchangers, where two fluids -- water and oil -- exchange heat without the two fluids mixing. Here it is heat transfer, *as the water* heats up by absorbing heat $dQ = \dot{m}\, c\, dT$ and the heat transferred in a distance dx is $dQ = h\, \pi\, D\, (T^* - T)\, dx$.

Again, once the system is successfully modeled, it is possible to change the system conditions, and in the problems section for this chapter the heat-transfer characteristics of a *counterflow* heat-exchanger are examined.

One may ask why differential equations are almost exclusively used in this text, and it is because a differential equation allows a system variable to be tracked in space and time. Using *Mathematica*, you may use **DSolve** to obtain a mathematical function, or **NDSolve** to find a numerical solution that by means of an *interpolation function* describes the system behavior. In either case, the criterion for good modeling of a dynamical system is whether the Physics and Mathematics are reasonably close to the behavior of the real physical system.

3.2 Track and Field

Following, is a list of world records for men in running, from 100-meters to 10,000 meters. The question before us is whether a Physics model of running can account for all the data.

Distance (m)	Time (s)	Avg Velocity (m/s)	Record Holder
100	9.70	10.28	Usain Bolt (JAM) 2008
200	19.32	10.35	Michael Johnson (USA) 1996
400	43.18	9.26	Michael Johnson (USA) 1999
800	101.1	7.91	Wilson Kipketer (DEN) 1997
1000	132.0	7.57	Noah Ngeny (KEN) 1999
1500	206.0	7.28	Hicham ElGuerrouji (MOR) 1999
1 mile	223.1	7.21	Hicham ElGuerrouji (MOR) 1999
2000	284.8	7.02	Hicham ElGuerrouji (MOR) 1999
3000	440.7	6.81	Daniel Komen (KEN) 1996
5000	757.4	6.60	Keninisa Bekelese (ETH) 2004
10000	1580.3	6.33	Keninisa Bekelese (ETH) 2004

Let us first of all put this data into a form where we can see the average velocity to run a race of a given distance. *Mathematica* can manipulate and plot data in an easy to read form.

```
vel = {{100, 10.28}, {200, 10.35}, {400, 9.26}, {800, 7.9}, {1000, 7.57},
    {1500, 7.28}, {1609, 7.21}, {2000, 7.02}, {3000, 6.81}, {5000, 6.6}}
velc = ListPlot[vel, PlotRange → {{-100, 5100}, {5, 11}}];
```

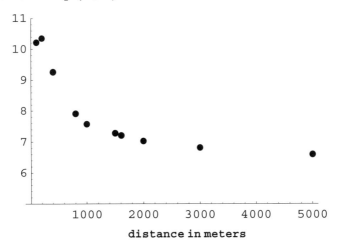

Let us first of all take an unbiased look at the data. For the short distance runs, 100-m and 200-m, the average velocities are very fast at over 10 m/s, diminishing almost exponentially to about 6 m/s as the run distance increases.

What this means is that the runner must go flat-out in a short distance run (the dash), and for a middle-distance or long-distance run, the runner must go at a slower pace so as not to burn up all the available energy.

Building the Differential Equations

Based on the above observations, let us model the acceleration phase of any race utilizing Newton's Law:

$$\frac{dv}{dt} = f - k v$$

f is the force per unit mass, and k is the resistance we encounter when we try and go faster.

We must also keep account of our energy

$$\frac{dE}{dt} = \sigma - f v$$

The energy in the runner's system per unit mass diminishes from an initial value E_o due to the power exerted in running $f \cdot v$ with an energy resupply rate of σ as the body begins to burn glucose, carbohydrates, and fat. These are converted into ATP which is the chemical source of muscular energy.

A Theory of Competitive Running

In September 1973, Joseph Keller published "A Theory of Competitive Running" in <u>Physics Today</u> (vol 26, pp 43-47). Keller utilized the above two equations with best-fit values of F, k, σ, and E_o and obtained a reasonable match with the world records of the time.

As it turns out, all the world records in Men's running have been broken over the last 30 years, so the purpose of this section is to revisit Keller's theory with a slightly-changed set of running parameters to see how closely the theory matches the world of 2008.

We choose the following physiological constants as reasonable for a male world-class runner

F = 1.25 * 9.8 m/s^2 (maximum acceleration of runner is 1.25 g)

k = 1.12 s^{-1} (internal and external resistance increases with velocity)

σ = 44.5 W/kg (rate of ATP resupply to muscles) [7% greater than Keller]

E_o= 2400 J/kg (initial energy and oxygen supply in muscles)

(These are the same constants chosen by J. Keller, except for σ, which is 7% larger than the 1973 value.)

We will now solve the first differential equation to find out how fast a world-class athlete could run the 100- and 200-meter dash. Then we will find out how far you can run at maximum speed before all the available energy is used up.

The Dash

For a short-distance run, the runner is accelerating all the way and maximum force F is exerted during the entire race. Then $\frac{dv}{dt} = F - k\,v$

Utilizing *Mathematica*,

```
In[67]:=  F = 1.25*9.8; k = 1.12; s = 44.5; E0 = 2400;
          sol =
          DSolve[{v'[t] == F - k*v[t], v[0] == 0}, v[t], t] // FullSimplify
          sol1 = DSolve[{x''[t] == F - k*x'[t], x[0] == x'[0] == 0}, x[t], t] //
          FullSimplify
          Plot[Evaluate[v[t] /. sol, {t, 0, 10}],
            PlotRange → {{0, 10.1}, {0, 11}},
            AxesLabel → {"Time (s)", "Velocity(m/s)"}];
```

$$Out[68]= \{\{v[t] \to 10.9375 - 10.9375\,e^{-1.12\,t}\}\}$$

$$Out[69]= \{\{x[t] \to -9.76563 + 9.76563\,e^{-1.12\,t} + 10.9375\,t\}\}$$

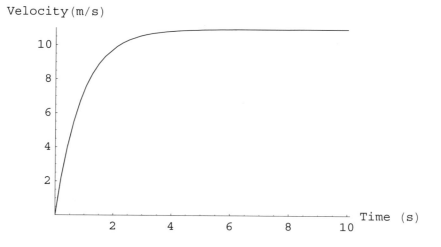

How long does it take us (according to the physics model) to do the 100- and 200-meter dash? We have the velocity with time, and assuming we accelerate the whole distance,

```
v[t]  =  10.9375*(1 - Exp[-1.12t])

x[t]  =  10.9375·t - 9.7656(1 - Exp[-1.12t])
```

Mathematica carries the solution **x[t]** in memory, so let us find the time at which our hypothetical runner crosses the 100-m and the 200-m line.

```
In[71]:=  sol100 = FindRoot[100 == x[t] /. sol1, {t, 10}]
          sol200 = FindRoot[200 == x[t] /. sol1, {t, 20}]

Out[71]=  {t → 10.0357}

Out[72]=  {t → 19.1786}
```

This compares favorably with the world record time 9.77 s in the 100-meter, and 19.32 s in the 200-meter dash.

How far can the runner keep this up, running as fast as possible, before running out of energy ?

Let's use the second equation

$$\frac{dE}{dt} = \sigma - F \cdot v \quad \text{and integrate to find} \quad E1 = F \cdot x1 - \sigma \cdot t$$

```
In[29]:=  Clear[x]; F = 1.25 * 9.8; k = 1.12; s = 44.5; E0 = 2400;
          x1[t_] = (F / k) * t - (F / k^2) * (1 - Exp[-k * t]);
          t1 = FindRoot[2400 == F * x1[t] - s * t, {t, 10}]
          Print["xmax = ", x1[t /. t1]]

Out[31]=  {t → 28.1572}

          xmax = 298.204
```

This is the furthest distance that a runner with the above physical constants can run, going flat-out. So the dash merely depends on the runners ability to accelerate and maintain maximum force throughout the run. *However*, if we want to run farther than 300 meters, then we will have to run *slower* (otherwise all the available energy is expended and we can go no further).

So here is the race strategy as given by Joseph Keller: accelerate in the first stage (time t) to velocity v1 and maintain that constant velocity throughout the remainder (time t2) of the race. Of course this is an oversimplification, but let us see if it produces reasonable results (one can always complicate the model later).

The Run

After the initial acceleration phase, the equations of motion are

$$\frac{dv}{dt} = f - k\,v = 0, \quad \text{or,} \quad f = k\,v$$

and, at the constant running speed, $\frac{dE}{dt} = \sigma - f\,v = \sigma - k\,v^2$

Notice that f is now much less than F, and with a reduced running speed, the rate of energy loss in the run-phase is diminished. So, for a run of distance $z > 300$ m, we accelerate for time t, reaching a velocity v1, and travel a distance x1, while burning energy E1. Then we maintain the constant velocity v1 for time t2.

We will solve for the time t2 at constant velocity, subject to the condition that all energy is used-up by the end of the race. Then, the total distance travelled is $z = x1 + x2 = x1 + v1 \cdot t2$ and the total time is $t + t2$.

The average velocity for running the race is then $w = \frac{z}{(t+t2)}$.

Also, for races less than 300 m, there is only the acceleration phase, and the distance travelled is x1 in time t. Writing the equations in general terms (so we can vary F, k, σ, and E_o if we wish),

```
F = 1.25 * 9.8; k = 1.12; s = 44.5; E0 = 2400;
x1[t_] = (F / k) * t - (F / k^2) * (1 - Exp[-k * t]) ;
v1[t_] = (F / k) * (1 - Exp[-k * t]);
E1[t_] = F * x1[t] - s * t ; E2[t_] = E0 - E1[t] ;


(* and since  E2 = (kv² - s) t2  *)
                E0 - E1[t]
t2[t_] = ───────────────────── ;
            (k * v1[t]^2 - s)
z[t_] = v1[t] * t2[t] + x1[t] ;
t3[t_] = t2[t] + t ;
w[t_] = z[t] / t3[t] ;
p = ParametricPlot[{z[t], w[t]},
    {t, 0.77, 4.2}, PlotRange → {{0, 5500}, {5, 11}}];
r = ParametricPlot[{x1[t], x1[t] / t}, {t, 3, 26.8},
    PlotRange → {{0, 5500}, {5, 11}}];
```

We will now use the above equations to find the times to run specific distances.

```
In[56]:=  (* -- Predicted times at various race distances-- *)
          sol400 = FindRoot[400 == v1[t] * t2[t] + x1[t], {t, 1}];
          t400 = (t2[t] + t) /. sol400;
          sol800 = FindRoot[800 == v1[t] * t2[t] + x1[t], {t, 1}] ;
          t800 = (t2[t] + t) /. sol800;
          sol1000 = FindRoot[1000 == v1[t] * t2[t] + x1[t], {t, 1}];
          t1000 = (t2[t] + t) /. sol1000;
          sol1500 = FindRoot[1500 == v1[t] * t2[t] + x1[t], {t, 1}] ;
          t1500 = (t2[t] + t) /. sol1500 ;
          sol1609 = FindRoot[1609 == v1[t] * t2[t] + x1[t], {t, 1}] ;
          t1609 = (t2[t] + t) /. sol1609;
          sol2000 = FindRoot[2000 == v1[t] * t2[t] + x1[t], {t, 1}] ;
          t2000 = (t2[t] + t) /. sol2000;
          sol3000 = FindRoot[3000 == v1[t] * t2[t] + x1[t], {t, 1}];
          t3000 = (t2[t] + t) /. sol3000;
          sol5000 = FindRoot[5000 == v1[t] * t2[t] + x1[t], {t, 1}] ;
          t5000 = (t2[t] + t) /. sol5000 ;
          sol10000 = FindRoot[10000 == v1[t] * t2[t] + x1[t], {t, 1}];
          t10000 = (t2[t] + t) /. sol10000;
          Print["  t100 = 10.035"]; Print ["  t200 = 19.178"];
          Print["  t400 = ", t400]; Print["  t800 = ", t800];
          Print[" t1000 = ", t1000];
          Print[" t1500 = ", t1500]; Print[" t1609 = ", t1609];
          Print[" t2000 = ", t2000]; Print[" t3000 = ", t3000];
          Print[" t5000 = ", t5000]; Print["t10000 = ", t10000]
```

	Theory
Race Dist	time(s)
t100 =	10.035
t200 =	19.178
t400 =	42.5421
t800 =	103.215
t1000 =	134.363
t1500 =	212.905
t1609 =	230.091
t2000 =	291.835
t3000 =	450.087
t5000 =	767.063
t10000 =	1560.06

Let's plot the theory line against the world record data points and see how we do.

```
Show[{velc, p, r}, AxesLabel → {"Distance(m)", "Velocity(m/s)"}]
```

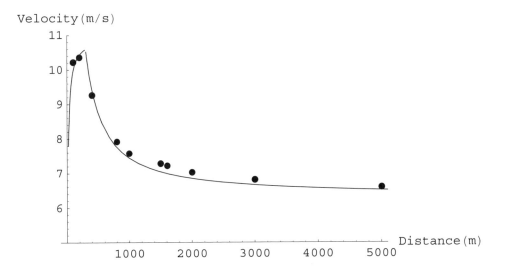

This appears to be a rather good fit, with the dots for the middle-distance runs actually exceeding the theory line. Only the 400 meter run is marginally below the theory line, which is already 3 % faster than the original theory of Joseph Keller. (Runners have certainly improved their racing performance dramatically over the last 30 years!) The last question is how well do the theoretical and actual record numbers compare?

Track and Field World Records -- Men 2008

Distance (m)	Theory Time	Record Time (s)	% Difference	
100	9.99	9.70	− 2.7	
200	19.11	19.32	1.1	
400	42.54	43.18	1.5	(record likely to fall)
800	103.2	101.1	− 2.1	
1000	134.4	132.0	− 1.9	
1500	212.9	206.0	− 3.4	
1 mile	230.1	223.1	− 3.2	
2000	291.8	284.8	− 2.6	
3000	450.1	440.7	− 2.2	
5000	767.1	757.4	− 1.4	
10000	1560.1	1580.3	1.2	

All the track records for men may be accounted for with the Physics model of running up to 10,000 meters. Beyond this range, when running fatigue sets in -- as the runner's system tries to clear the lactic acid build-up, acidosis of the muscles begins to occur -- this is often referred to as THE WALL. Many long-distance runners have learned to cope with the fatigue by training and diet, and also by running at a somewhat reduced pace. This can be accounted for in the Physics model by allowing k (the internal resistance) to increase for runs beyond 10,000 meters. For track and field records for women, see Challenge problem 3-6.

Technically, the supply of ATP to the runner's muscles is much more compli- cated than just a constant number. The biochemistry of running is carefully discussed in Henry Bent's article "Energy and Exercise" in the *Journal of Chemi- cal Education*, vol 55, p 797 (1978), and in even greater detail by Jack Wilmore and David Costill (2004) *Physiology of Sport and Exercise* [Human Kinetics publications].

A capsule summary is as follows: In the first 3-5 seconds, almost all the ATP is "burned-up" ATP→ ADP + Energy. Then, over the next 10-20 seconds, the "burned" ATP is reconstituted by creatine phosphate   ADP + CP → ATP . This is all the "ready-reserve" in the runner's system, and after this, the runner burns glycogen, first anaerobically and then aerobically to supply ATP to the muscles on a continuous basis.

A more complete (and complicated) theory of competitive running would factor in the non- constant rate of energy resupply to the runner.

Note added for Expanded Edition (2009):

Usain Bolt (Lightning Bolt) of Jamaica has now reset the world record for the 100-m and 200-m distances (100-m 9.58 s, 200-m 19.19 s). We may account for Bolt's excellent times by increasing the maximum acceleration in the 100 meter dash by 5 %. Then F =1.312 * 9.8 m/s^2 and t→ 9.60 s.

A very interesting physics paper would be to analyze a videotape of Bolt's record-breaking runs in Berlin in August 2009, and determine his velocity and acceleration throughout the race.

3.3 The Binary Pulsar 1913 + 16

Some 20,000 lightyears from the Earth is an amazing object: two neutron stars (or *pulsars*) locked in an orbit so tight that the two stars complete an orbit in just 7.75 hours. The mass of each star is very nearly equal to 1.4 M_{sun}, and astronomers have determined the eccentricity of the orbit is $\epsilon = 0.617$.

What we would like to do with *Mathematica* is the following: First, find the "semi-major" axis, the distance between the centers of the two stars' orbits. Then do a basic polar plot of the orbit.

Second, solve the equations of motion, where Star1 is at (x,y) and Star 2 is at (-x,-y). Then we will Table the results, and finally Animate the plot.

The Physics of Binary Stars

When two stars are locked in a binary system, the parameters that describe the system are

a   the semi-major axis, which is the <u>average</u> of the periastron (closest) and apastron (furthest) separations of the two stars. $a = \frac{1}{2}(d_{peri} + d_{Ap})$

ϵ the orbital eccentricity, which is the <u>elongation</u> of the orbit of either star. The closer ϵ is to one, the further the orbit is away from circular.

v_p the periastron velocity $v_p = \frac{2 \pi r}{T} \sqrt{\frac{1 + \epsilon}{1 - \epsilon}}$

v_A the apastron velocity $v_A = \frac{2 \pi r}{T} \sqrt{\frac{1 - \epsilon}{1 + \epsilon}}$

where r is the *average radius* of the star's orbit measured from the CM

d_p the periastron separation of the stars $d_p = a (1 - \epsilon)$

d_A the apastron separation of the stars $d_A = a (1 + \epsilon)$

We have almost everything we need, for all these quantities, except a which we will now find by using the Newton-Kepler Law

$$T^2 = \frac{4 \pi^2 a^3}{G (M1 + M2)}$$

```
In[25]:=  Clear[x, v, a, t];  T = 7.75 * 3600;  M1 = M2 = 1.4 * 2 * 10^30;
          G = 6.67 * 10^-20;   (* distances in km, time in seconds*)
          FindRoot[T^2 == 4 π^2 * a^3 / (G * 2 M1), {a, 10^6}]
```

$Out[28] = \{a \to 1.9456 \times 10^6\}$

```
In[15]:=  dp = 1.945 * 10^6 * (1 - .617)          da = 1.945 * 10^6 * (1 + .617)
```

$Out[15] =$ 744935. 3.14507×10^6

$In[17] := \mathbf{a = 1.9456 * 10^6; \ \epsilon = .617; \ T = 7.75 * 3600; \ r = a / 2;}$

$$\mathbf{vp = \frac{2 \pi r}{T} \sqrt{\frac{(1 + \epsilon)}{(1 - \epsilon)}}} \qquad\qquad \mathbf{va = \frac{2 \pi r}{T} \sqrt{\frac{(1 - \epsilon)}{(1 + \epsilon)}}}$$

$Out[18] = \qquad\qquad 450.148 \qquad\qquad\qquad 106.621$

We now know everything about the orbits of the two stars. We will now polar plot the orbits of the two stars. For each star, $r = \dfrac{a \ (1 - \epsilon^2)}{1 + \epsilon \cos \theta}$

```
In[1]:=  Needs["Graphics`Graphics`"];
         a = (1.945 * 10^6) / 2; ε = .617;
         stars1 = PolarPlot[a * (1 - ε^2) / (1 + ε * Cos[θ]),
              {θ, 0, 2 π}, AspectRatio → 0.5];
         stars2 = PolarPlot[a * (1 - ε^2) / (1 - ε * Cos[θ]),
              {θ, 0, 2 π}, AspectRatio → 0.5];
         Show[{stars1, stars2}, AspectRatio → 0.3]
```

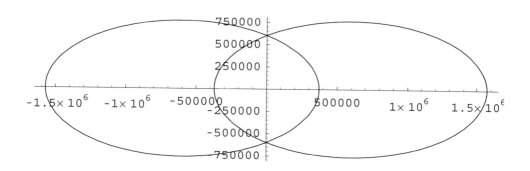

The two stars approach within 744,000 km of each other at periastron (the point of closest approach.) At this time, each star is moving at 450 km/s. The two stars move past each other at periastron like two bullet trains moving in opposite directions at one million miles per hour. The bending of space is so great at this distance (which is only half the diameter of our sun) that the binary system emits *gravitational waves,* and *with this loss of energy, the orbit of the two stars is slowly collapsing.* (See problem 1 in Chapter 3 for the length of time for this binary system to collapse into a black hole.)

Due to the *symmetry* of this problem, with the masses of the two stars equal, we may write the differential equation of motion of Star 1 to apply at (x,y) with the understanding that Star 2 will be at (-x, -y). Therefore we only have to solve one set of Differential Equations. (The case of stars of unequal masses will be treated in Section 3.4).

Building the Differential Equation

The gravitational forces acting on Star 1 at (x, y) from Star 2 at $(-x, -y)$ are

$$Mx'' = -\frac{MGM}{(2x)^2 + (2y)^2}\frac{2x}{\sqrt{(2x)^2 + (2y)^2}} \qquad x_o = 1.572 \times 10^6 \text{ km} \qquad \dot{x}_o = 0$$

$$My'' = -\frac{MGM}{(2x)^2 + (2y)^2}\frac{2y}{\sqrt{(2x)^2 + (2y)^2}} \qquad y_o = 0 \qquad \dot{y}_o = 106.6 \text{ km/s}$$

Notice that the gravitational attraction between the two stars is MGM/r^2 where the distance r between the two stars is $\sqrt{(2x)^2 + (2y)^2}$ and the Cosine of the angle from the CM to Star 1 is $2x/r$. We will start the two stars at their Apastron distance, each 1.57×10^6 km from the CM.

Programming in *Mathematica*, with distance in km and time in seconds,

```
Clear[x, y, vel, t] ;
sol =
  NDSolve[{x''[t] == -60^2 * 2.8 * 10^30 * 6.67 * 10^-20 * (2 x[t]) /
      (4 x[t]^2 + 4 y[t]^2)^1.5, y''[t] == 60^2 * 2.8 * 10^30 *
      6.67 * 10^-20 * (-2 y[t]) / (4 x[t]^2 + 4 y[t]^2)^1.5,
    x[0] == 1.5725 * 10^6, y[0] == 0, x'[0] == 0, y'[0] == 106.6 * 60},
    {x, y}, {t, 0, 500}]
InterpFunc1 = x /. sol[[1]]; InterpFunc2 = y /. sol[[1]];
InterpFunc3 = x' /. sol[[1]]; InterpFunc4 = y' /. sol[[1]];
sep[t_] = 2 * √InterpFunc1[t]^2 + InterpFunc2[t]^2 ;
vel[t_] = √InterpFunc3[t]^2 + InterpFunc4[t]^2 ;
Print["max velocity=", vel[232.5] / 60,
  " km/s   separation =", sep[232.5], " km"]

T2 = Table[{t, InterpFunc1[t], Chop[InterpFunc2[t]],
    Chop[InterpFunc3[t] / 60], InterpFunc4[t] / 60, vel[t] / 60},
    {t, 0, 465, 15}] // TableForm
```

From In[344]:= max velocity = 450.464 km/s   separation = 744250. km

Table of values for the binary pulsar system. The orbital period is 465 min. The maximum velocity of Star 1 and Star 2 is 450 km/s (1 million mph) at a distance of only 744,000 km.

t (min)	x (km)	y (km)	x' (km / s)	y' (km / s)	v (km / s)
0	1.572×10^6	0	0	106.6	106.6
15	1.564×10^6	95784.	-17.	106.0	107.4
30	1.541×10^6	190625.	-34.	104.4	109.9
45	1.503×10^6	283535.	-51.	101.7	114.1
60	1.448×10^6	373438.	-69.	97.78	119.9
75	1.377×10^6	459104.	-88.	92.31	127.5
90	1.290×10^6	539076.	-107	85.06	137.
105	1.184×10^6	611545.	-127	75.55	148.4
120	1.059×10^6	674181.	-149	63.06	162.2
135	914677.	723841.	-172	46.48	178.9
150	747933.	756077.	-198	23.95	199.4
165	557697.	764229.	-224	-7.74	225.1
180	342719.	737653.	-252	-54.5	258.4
195	104400.	658041.	-275	-128.	303.5
210	-144018.	492294.	-267	-250.	366.1
225	-340834.	192498.	-136	-414.	436.5
240	-337889.	-201197	142.	-411.	435.2
255	-138375.	-497527	268.	-246.	364.4
270	110193.	-660721	274.	-126.	302.3
285	348035.	-738789	251.	-53.2	257.5
300	562431.	-764383	224.	-6.85	224.4
315	752099.	-755566	197.	24.56	198.9
330	918313.	-722857	172.	46.92	178.5
345	1.062×10^6	-672848	148.	63.4	161.9
360	1.186×10^6	-609950	127.	75.80	148.1
375	1.292×10^6	-537281	106.	85.25	136.7
390	1.379×10^6	-457157	87.6	92.46	127.3
405	1.450×10^6	-371376	69.0	97.89	119.8
420	1.504×10^6	-281390	51.2	101.8	114.0
435	1.542×10^6	-188422	33.7	104.5	109.8
450	1.565×10^6	-93548.	16.6	106.1	107.3
465	1.572×10^6	2246.4	-0.3	106.6	106.6

To better represent the orbits of the pulsars as they move about the CM, an Animated series of plots is given by the following **Do Loop.** A dozen stills from this Animation are shown. To see the entire Animation, access the program *BinaryStarsOrbit* on the CD, and run it in *Mathematica.*

```
In[27]:= g1 = ParametricPlot[{Table[{InterpFunc1[t], InterpFunc2[t]}],
            Table[{-InterpFunc1[t], -InterpFunc2[t]}]}, {t, 0, 465},
            AspectRatio → Automatic, PlotStyle → GrayLevel[0.5],
            PlotRange → {{-2 * 10^6, 2 * 10^6}, {-1 * 10^6, 1 * 10^6}},
            AxesStyle → GrayLevel[0.65], Ticks → None];
```

```
In[28]:= Do[(g2 = ListPlot[{{InterpFunc1[i], InterpFunc2[i]},
            {-InterpFunc1[i], -InterpFunc2[i]}},
            PlotRange → {{-2 * 10^6, 2 * 10^6}, {-1 * 10^6, 1 * 10^6}},
            DisplayFunction → Identity, PlotStyle → PointSize[0.03]];
            Show[g1, g2];), {i, 0, 437, 23}]
```

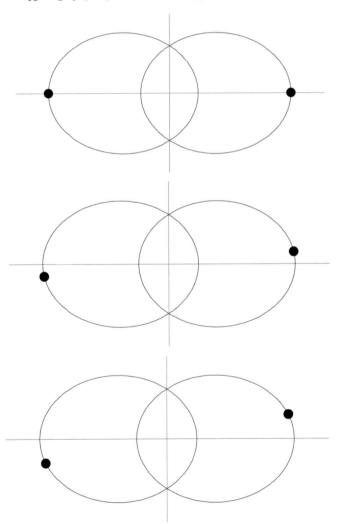

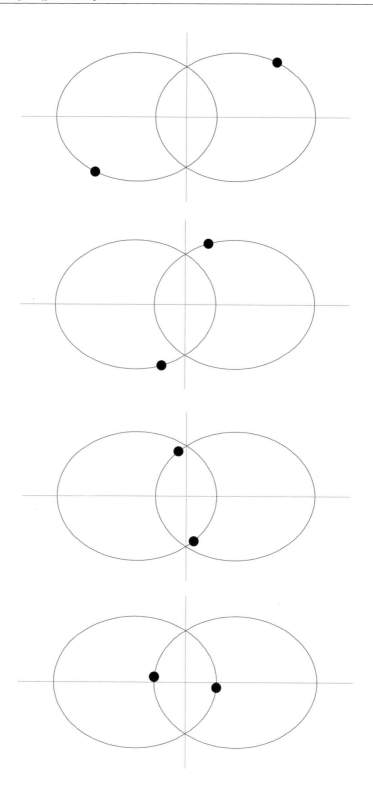

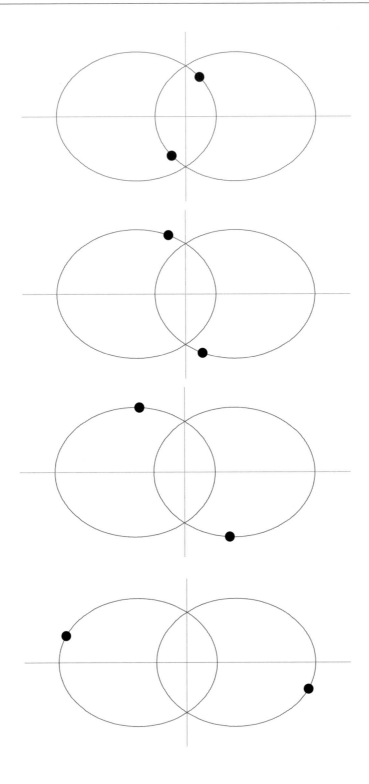

3.4 Exploding Star in a Binary System

A 15 M_{sun} star is in a binary system with an 8 M_{sun} star. The stars are 1.5×10^8 km apart, in circular orbits about their Center-of-Mass. Suddenly, the 15 M_{sun} star goes supernova in a spherically symmetric explosion, reducing its mass to 3 M_{sun}. What are the velocities of the stars after the explosion ? Does the system become unbound ?

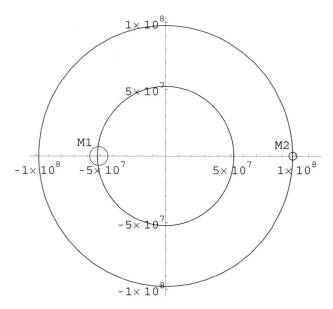

Before the explosion of M1, the two stars are in circular orbits about the CM. Let's first of all find the orbit radius a of star M1 and orbit radius b of M2. Then use the Newton-Kepler Law to find the orbit time T of each star, then find V1 and V2.

```
In[55]:=   M1 = 15; M2 = 8; r = 1.5 * 10^8;
           sol2 = Solve[{M1 * a == M2 * b, r == a + b}, {a, b}]
```

```
Out[56]=   {{a → 5.21739 × 10^7, b → 9.78261 × 10^7}}
```

```
In[57]:=   G = 6.67 * 10^-20;  (* x and y in km, T in seconds*)
           FindRoot[T^2 == 4 π^2 * r^3 / (G * (M1 + M2) * 2 * 10^30), {T, 10^6}]
```

```
Out[58]=   {T → 6.58984 × 10^6}
```

```
           T = 6.5898 * 10^6; a = 5.22 * 10^7; b = 9.78 * 10^7;
           v1 = 2 π * a / T; v2 = 2 π * b / T; Print["v1→", v1, "  v2→", v2 ]

               v1→49.7712   v2→93.2495 km/s
```

Building the Differential Equation

We are now ready to write the differential equations of the two stars. Since both stars orbit the center-of-mass at $(0,0)$ we will call the coordinates of star M1 (x,y) and the coordinates of star M2 (w,z). Then the accelerations of the two stars are

$$x'' = -M2\ G\ \frac{(x-w)}{\left((x-w)^2 + (y-z)^2\right)^{3/2}} \qquad x_o = -5.22 \times 10^7 \quad x_o' = 0$$

$$y'' = -M2\ G\ \frac{(y-z)}{\left((x-w)^2 + (y-z)^2\right)^{3/2}} \qquad y_o = 0, \qquad y_o' = -49.8$$

$$w'' = -M1^*\ G\ \frac{(w-x)}{\left((x-w)^2 + (y-z)^2\right)^{3/2}} \qquad w_o = +9.78 \times 10^7 \quad w_o' = 0$$

$$z'' = -M1^*\ G\ \frac{(z-y)}{\left((x-w)^2 + (y-z)^2\right)^{3/2}} \qquad z_o = 0, \qquad z_o' = +93.3$$

* Where M1 is 15 M_{Sun} before $t = 0$, and 3 M_{Sun} after $t = 0$.

The positions of the two stars are shown from 960 hours before the explosion up to the moment of DETONATION $(t = 0)$. It is at this time that Star M1 explodes.

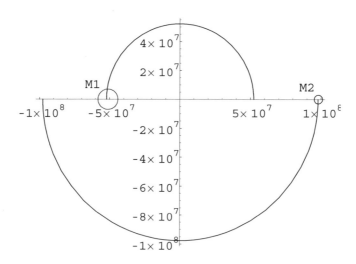

The subsequent paths of the two stars is determined by their gravitational interaction. The mass of M1 after the explosion is 3 M_{sun}. M2 remains at 8 M_{sun}. The programming in *Mathematica* proceeds as follows:

```
In[109]:=
    (* distance in km, time in hr *)
    Clear[x, y, v, t]
    sol = NDSolve[{x''[t] == -3600^2 * 16 * 6.67 * 10^10 *
        (x[t] - w[t]) / ((x[t] - w[t])^2 + (y[t] - z[t])^2)^1.5,
      y''[t] == -3600^2 * 16 * 6.67 * 10^10 *
        (y[t] - z[t]) / ((x[t] - w[t])^2 + (y[t] - z[t])^2)^1.5,
      w''[t] == -3600^2 * 30 * If[t < 0, 1, .2] * 6.67 * 10^10 *
        (w[t] - x[t]) / ((x[t] - w[t])^2 + (y[t] - z[t])^2)^1.5,
      z''[t] == -3600^2 * 30 * If[t < 0, 1, .2] * 6.67 * 10^10 *
        (z[t] - y[t]) / ((x[t] - w[t])^2 + (y[t] - z[t])^2)^1.5,
      x[0] == -5.22 * 10^7, y[0] == 0, w[0] == 9.78 * 10^7, z[0] == 0,
      x'[0] == 0, y'[0] == -49.8 * 3600, w'[0] == 0, z'[0] == 93.3 * 3600},
     {x, y, w, z}, {t, -960, 1960}]
    InterpFunc1 = x /. sol[[1]]; InterpFunc2 = y /. sol[[1]];
    InterpFunc3 = x' /. sol[[1]]; InterpFunc4 = y' /. sol[[1]];
    InterpFunc5 = w /. sol[[1]]; InterpFunc6 = z /. sol[[1]];
    InterpFunc7 = w' /. sol[[1]]; InterpFunc8 = z' /. sol[[1]];
    rad[t_] = √((InterpFunc1[t] - InterpFunc5[t])^2 +
        (InterpFunc2[t] - InterpFunc6[t])^2);
    vel1[t_] = √(InterpFunc3[t]^2 + InterpFunc4[t]^2);
    vel2[t_] = √(InterpFunc7[t]^2 + InterpFunc8[t]^2);
    xcm[t_] =
      (3 * InterpFunc1[t] + 8 * InterpFunc5[t]) / 11 * If[t < 0, 0, 1];
    ycm[t_] = (3 * InterpFunc2[t] + 8 * InterpFunc6[t]) / 11 *
        If[t < 0, 0, 1];
    ParametricPlot[{{x[t], y[t]} /. sol, {w[t], z[t]} /. sol},
      {t, -912, 1500}, AspectRatio → Automatic,
      Prolog → {{Text["M1", {-6.2 * 10^7, 10^7}],
          Hue[.91], Circle[{-5.12 * 10^7, 10}, 7000000]},
        {Text["M2", {11 * 10^7, .8 * 10^7}], Hue[0.67],
          Circle[{9.8 * 10^7, 100}, 3000000]}}];
    T2 = Table[{t, rad[t] / 10^6, vel1[t] / 3600,
        vel2[t] / 3600, xcm[t] / 10^6, ycm[t] / 10^6},
      {t, -168, 600, 24}] // TableForm
```

The **IF** command allows us to plot the evolution of the orbit from 7 days before the explosion (where the orbits are circular) to 25 days after the explosion (as the orbits diverge). That's it. *Mathematica* easily handles 4 simultaneous differential equations.

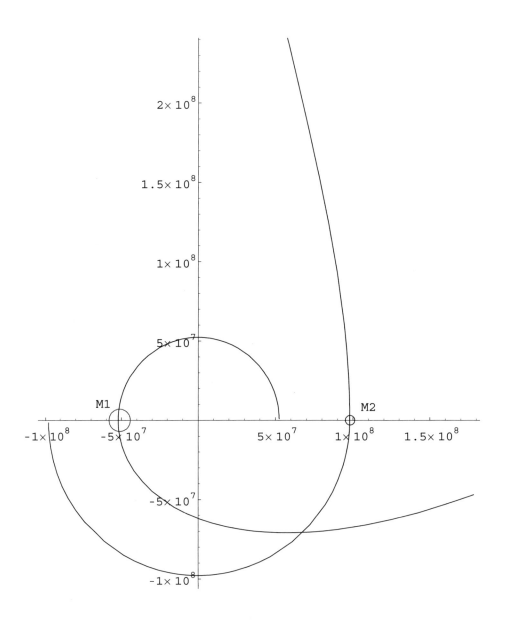

M2 tears off into space at high velocity. The binary system is now unbound.

Time and Distance between stars in 10^6 kilometers, Velocities, Location of CM in 10^6 km

t (hr)	r (10^6 km)	v1 (km / s)	v2 (km / s)	CM – X	CM – Y
-168	150.027	49.7868	93.2873	0	0
-144	150.02	49.7902	93.2906	0	0
-120	150.014	49.7932	93.2934	0	0
-96	150.009	49.7956	93.2958	0	0
-72	150.005	49.7975	93.2976	0	0
-48	150.002	49.7989	93.2989	0	0
-24	150.001	49.7997	93.2997	0	0
0	150.	49.8	93.3	56.8909	0.
24	150.266	49.7997	93.2495	56.8909	4.68916
48	151.06	49.7989	93.0994	56.8909	9.37833
72	152.371	49.7976	92.8545	56.8909	14.0675
96	154.182	49.7958	92.522	56.8909	18.7567
120	156.47	49.7935	92.1113	56.8909	23.4458
144	159.208	49.7909	91.633	56.8909	28.135
168	162.367	49.7881	91.0984	56.8909	32.8241
192	165.913	49.785	90.5189	56.8909	37.5133
216	169.815	49.7817	89.9052	56.8909	42.2025
240	174.039	49.7784	89.2674	56.8909	46.8916
264	178.554	49.775	88.6144	56.8909	51.5808
288	183.329	49.7715	87.9538	56.8909	56.27
312	188.338	49.7681	87.292	56.8909	60.9591
336	193.552	49.7648	86.6344	56.8909	65.6483
360	198.949	49.7615	85.9853	56.8909	70.3375
384	204.506	49.7583	85.348	56.8909	75.0266
408	210.203	49.7552	84.7249	56.8909	79.7158
432	216.023	49.7522	84.118	56.8909	84.4049
456	221.949	49.7493	83.5286	56.8909	89.0941
480	227.968	49.7465	82.9573	56.8909	93.7833
504	234.066	49.7438	82.4047	56.8909	98.4724
528	240.233	49.7412	81.871	56.8909	103.162
552	246.458	49.7387	81.356	56.8909	107.851
576	252.732	49.7364	80.8595	56.8909	112.54
600	259.049	49.7341	80.3811	56.8909	117.229

In[128]:= **vy_cm = 4.689 * 10^6 / 24 / 3600**

54.2708

An extremely interesting result from the above table is that the Center-of-Mass of the binary system shifts from (0,0) to (56.8×10^6,0) right after detonation. Then the CM moves vertically upward at 54 km/s as the stars separate.

For more on exploding stars in binary systems, see Romas Mitalas *Am J Physics* (Mar 1980) "Supernovae in Binary Systems" pp 226-231.

3.5   Heat Exchanger

Before we get to the very interesting subject of the heat-exchanger, let us take a somewhat simpler case of extracting heat from a hot environment. Say we start with water flowing through a tube of diameter D and total length L being used to extract heat from an operating automobile engine.

If water flowing at 0.8 m/s enters a 4-cm diameter tube at 25 °C and the walls of the tube inside the engine are maintained at 100 °C, how much is the water heated in passing 4-meters total distance through the engine? And how much heat is extracted per second?

To solve this problem, we need to set-up a basic differential equation, where the heat extracted from the engine block as the fluid flows a distance dx is $dQ = h \, \pi D \, dx \, (T^* - T)$ and the water absorbs dQ and heats up by an amount dT $dQ = \dot{m} \, c \, dT$

Setting these two equations equal, we have our basic differential equation

$$\frac{dT}{dx} = \frac{h \, \pi \, D \, (T^* - T)}{\dot{m} \, c} \quad \text{(Eq 1)}$$

where $T^* = 100$ °C, $D = 0.04$ m, $c = 4180$ J/kg °C, $\dot{m} = 1$ kg/s, $L = 4$ m, and $T_o = 25$ °C.

Before solving, we first have to find h, the convective heat transfer coefficient. This depends on the heat transfer characteristics of water, and is given in the engineering literature by the *Petukhov equation*

$$h = \frac{k}{D} \, \frac{(f/8) \, (RN - 1000) \, PR}{1 + 12.7 \, (f/8)^{.50} \, (PR^{2/3} - 1)} \quad \text{for } 3000 < RN < 5 \times 10^6 \text{ and } PR > 0.5$$

where RN is Reynolds number, PR is the Prandtl number and

the friction factor $f = (0.79 \, \text{Ln} \, RN - 1.64)^{-2}$

For water at 25 °C, $\nu = 0.9 \times 10^{-6}$ m²/ s, PR = 6.0, and k = 0.612 W/m °C

Thus $RN = \frac{.8 * .04}{.9 \times 10^{-6}} = 35{,}550$ (turbulent flow) and $f = 0.0227$

Now, solving the above Eq 1 for T(x)

```
DSolve[{T'[x] == h * π D / mc * (100 - T[x]), T[0] == 25}, T[x], x] //
FullSimplify
```

$Out[23] = \left\{ \left\{ T[x] \rightarrow 100 - 75 \, e^{-\frac{D \, h \, \pi \, x}{m \, c}} \right\} \right\}$

We now have the form of T(x), but we still need to find the heat transfer coefficient h

In[10]:= **k = .612; D1 = .04; PR = 6.0; RN = 35550; f = .0227;**

$$h = \frac{k}{D1} \frac{(f/8) * (RN - 1000) * PR}{1 + 12.7\sqrt{f/8}\ (PR^{2/3} - 1)}$$

Out[11]= 3519.76

In[12]:= **Clear[T, t]; mc = 1 * 4180; h = 3519.76; D1 = .04;**
sol =
DSolve[{T'[x] == h * π * D1 / mc * (100 - T[x]), T[0] == 25}, T[x], x]
Plot[Evaluate[T[x] /. sol], {x, 0, 10}, PlotRange → {0, 100}]
T[x_] = T[x] /. sol;
T[4]

Out[13]= {{T[x] → 100. − 75. e$^{-0.105815\,x}$}}

Out[16]= {50.8819}

The heat transferred per second is $\dot{Q} = \dot{m}\,c\,\Delta T = 1 * 4180 * 25.88 = 108$ kW

A Parallel-flow Heat Exchanger

A heat exchanger is a device for transferring heat from one fluid to another without mixing. In a *parallel-flow* heat exchanger, the cold fluid enters from the left and exits heated on the right, and the hot fluid also enters from the left and exits after cooling on the right. If hot oil at 200 °C with a mass flow rate of 3 kg/s and a specific heat of 1900 J/kg ·°C enters the outer shell from the left, and cool water at 20 °C with a mass flow rate of 0.9 kg/s and specific heat 4180 J/kg ·°C enters the inner tube from the left, what is the rate of heat exchange in this heat exchanger? We will assume a heat-transfer coefficient U = 460 W/m^2·°C, and the heat-exchange surface area is $A = \pi\,D\,L$. Where D = 10 cm, and length L = 40 m.

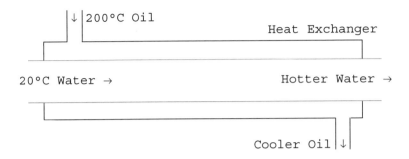

A parallel-flow Heat Exchanger

Building the Differential Equations

For a heat exchanger, the cold fluid is heated dT by absorbing heat energy dQ through an area π D dx.

In the inner tube,

$$dQ = (\dot{m}\,c)_c \, dT = U \, \pi \, D \, (T^* - T) \, dx \qquad \text{Cold side}$$
$$\text{Heat Absorbed} = \text{Heat Transferred}$$

In the outer shell, the hot fluid (which is moving along + dx) *loses* heat energy dQ through the same area.

In the outer shell,

$$dQ = (\dot{m}\,c)_h \, dT^* = - U \, \pi \, D \, (T^* - T) \, (dx) \qquad \text{Hot side}$$
$$\text{Heat Lost} = \text{Heat Transferred}$$

When we write the differential equations, with T for the cold fluid and T^* for the hot fluid,

$$(\dot{m}\,c)_c \, \frac{dT}{dx} = U \, \pi \, D \, (T^* - T) \qquad \text{and}$$
$$(\dot{m}\,c)_h \, \frac{dT^*}{dx} = U \, \pi \, D \, (T^* - T)$$

Now, programming in *Mathematica*,

```
In[155]:=
    mh = 3.0 * 1900; mc = 0.9 * 4180; D0 = .10; L = 40; U = 460; A = π * D0 * L;
    sol = NDSolve[{mc * T'[x] == U * π * D0 * (S[x] - T[x]),
       mh * S'[x] == - U * π * D0 * (S[x] - T[x]),
       T[0] == 20, S[0] == 200}, {S, T}, {x, 0, 80.0}]
    InterpFunc1 = S /. sol[[1]]; InterpFunc2 = T /. sol[[1]];
    Table[{x, Chop[InterpFunc1[x]], Chop[InterpFunc2[x]]},
       {x, 0, 40, 5}] // TableForm
    Plot[{S[x] /. sol, T[x] /. sol}, {x, 0, 40},
       AxesLabel → {"Dist(m)", "Temp(°C)"},
       Epilog → {Text["Oil", {17, 168}], Text["Water", {27, 85}]}];
```

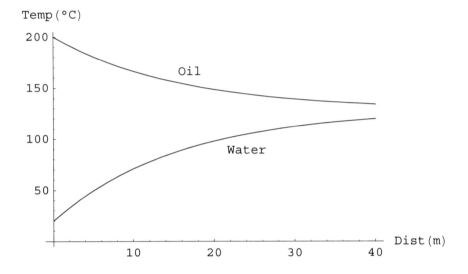

The Parallel-flow heat exchanger

x (m)	S (hot)	T (cold)
0	200.	20.
5	180.462	49.603
10	166.258	71.1242
15	155.932	86.77
20	148.425	98.1444
25	142.967	106.414
30	138.999	112.425
35	136.115	116.796
40	134.018	119.973

```
Plot[4, {x, -12, 10.5}, PlotRange → {-0.91, .6}, Axes → None,
 Prolog → {Text["134°C Oil ↓", {5.65, -.67}],
   Text["↓ 200°C Oil ", {-6.42, 0.15}],
   Text["Heat Exchanger", {6.8, .07}], Text["120°C Water →",
    {7.78, -.25}], Text["20°C Water →", {-9.7, -.25}],
   Line[{{-11.0, -.5}, {7.59, -.5}}], Line[{{8.58, -.5}, {9.3, -.5}}],
   Line[{{7.6, -.5}, {7.6, -.7}}], Line[{{8.5, -.5}, {8.5, -.7}}],
   Line[{{-9.5, .2}, {-9.5, 0}}], Line[{{-8.5, .2}, {-8.5, 0}}],
   Line[{{9.3, -.12}, {9.3, 0}}], Line[{{9.3, -.5}, {9.3, -.39}}],
   Line[{{-11, -.12}, {-11, 0}}], Line[{{-11, -.5}, {-11, -.39}}],
   Line[{{-11.0, 0}, {-9.5, 0}}], Line[{{-8.51, 0}, {9.3, 0}}]},
 Epilog → {GrayLevel[0.6], Line[{{-12, -.39}, {10.5, -.39}}],
   Line[{{-12, -.120}, {10.5, -.120}}]}]]
```

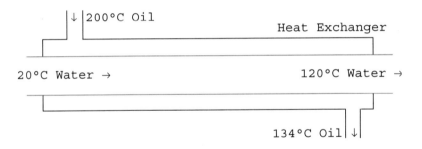

A Parallel-flow Heat Exchanger

The hot oil is cooled from 200 °C to 134 °C
while the water is heated from 20 °C to 120 °C.
Only heat is exchanged. The fluids do not mix.

3.6 Challenge Problems

1. Gravity Waves and Pulsar 1913+16

In 1964, ten years before the discovery of the binary pulsar, Philip C. Peters published a paper "Gravitational Radiation from Two Point Masses" (*Physical Review* 136, B1224-1232).

In this paper, the author derives a series of results from general relativity theory. Let us solve each of the following for the binary pulsar ($M1 = M2 = 2.8 \times 10^{30}$ kg, $\epsilon = 0.617$, $P_o = 27900$ s, $a = 1.95 \times 10^9$m, $c = 3 \times 10^8$m/s, $G = 6.67 \times 10^{-11}$ m^3/kg·s^2)

A. Rate of Energy Loss due to Gravitational Radiation

$$\frac{dE}{dt} = -\frac{32}{5} \frac{G^4 \, M1^2 \cdot M2^2 \, (M1 + M2)}{c^5 \cdot a^5 \, (1-\epsilon^2)^{7/2}} (1 + \tfrac{73}{24} \epsilon^2 + \tfrac{37}{96} \epsilon^4)$$

B. Rate of Decrease of "Semi-major" axis of Binary System

$$\frac{da}{dt} = -\frac{64}{5} \frac{G^3 \, M1 \cdot M2 \, (M1 + M2)}{c^5 \cdot a^3 \, (1-\epsilon^2)^{7/2}} (1 + \tfrac{73}{24} \epsilon^2 + \tfrac{37}{96} \epsilon^4)$$

C. Rate at which the Orbital Period decreases

$$\frac{dP}{dt} = -\frac{96}{5} \frac{G^3 \, M1 \cdot M2 \, (M1 + M2)}{c^5 \cdot P^{5/3} \, (1-\epsilon^2)^{7/2}} \left(\frac{4 \, \pi^2}{G \, (M1+M2)} \right)^{4/3} (1 + \tfrac{73}{24} \epsilon^2 + \tfrac{37}{96} \epsilon^4)$$

D. Time (in s) for the two stars to spiral in (and become a black hole)

$$T_{collapse} = \frac{12}{19} \frac{\left(1.55 \times 10^9\right)^4}{6.86 \times 10^{19}} \int_0^{.617} \epsilon^{29/19} \frac{\left(1 + \tfrac{121}{304} \epsilon^2\right)^{.5137}}{(1-\epsilon^2)^{1.5}} \, d\epsilon$$

Astronomers have observed that the orbital period of the binary pulsar decreases at 76.5 μ sec/ year. Using the Newton–Kepler Law, this results in a decrease in the semi-major axis of 3.5 m/year.

Solve the above equations of General Relativity to find out how much of the decrease in the orbital period and in the semi-major axis is due to the emission of gravity waves.

2. Halley's Comet Equations

Use the equations of Section 3.3 to find all the orbital parameters of Halley's comet, given only its velocity of 54.45 km/s at closest approach to the Sun, which was 0.59 AU in 1986. The mass of the Sun is 2×10^{30} kg.

Find orbital time T, eccentricity ϵ, semi-major axis a, and maximum and minimum velocity and distance of the comet from the Sun. Then polar plot Halley's orbit to scale.

3. The Largest Stars

Plaskett's binary star system consists of two stars that revolve in a circular orbit about a center of gravity midway between them. This means the masses of the two stars are equal.

If the orbital velocity of each star is 220 km/s and the orbital period of each is 14.4 days, find the mass M of each star.

4. The Alaska Pipeline

The Alaska pipeline is 800 miles long, stretching from Prudhoe Bay in the North to the port of Valdez in the South. About 1 million barrels of oil per day is pumped through the 1.2-meter diameter pipeline (accounting for 5% of the United States daily consumption).

The oil in the pipeline is hot, at 50°C, and to prevent heat loss, the pipeline is wrapped with 10-cm of fiberglass insulation of conductivity k = 0.035 W/m·°C.

For the above-ground pipeline, find the heat-loss per meter and the temperature of the surface of the fiberglass-wrapped pipeline if, on a cold winter's night, the air temperature is -40 °C, and the convective heat transfer coefficient is h = 12 W/ m^2·°C.

5. Counterflow Heat Exchanger

In a *counterflow* heat exchanger, the cold fluid enters from the left and exits heated on the right, and the hot fluid enters from the right and exits after cooling on the left. If hot oil at 200 °C with a mass flow rate of 3 kg/s and a specific heat of 1900 J/kg ·°C enters the outer shell from the right, and cool water at 20 °C with a mass flow rate of 0.9 kg/s and specific heat 4180 J/kg ·°C enters the inner tube from the left, what is the rate of heat exchange in this heat exchanger? We will keep the same heat-exchange parameters as in Section 3-5, the heat-transfer coefficient U = 460 W/m^2·°C, and the heat-exchange surface area is A = π D L. Where D =10 cm, and the length is L = 40 m.

Be sure to compare the effectiveness of this counter-flow heat exchanger with the effectiveness of the parallel-flow heat exchanger of Section 3-5. Which system, *parallel* or *counterflow*, does the better job of extracting heat from the hot fluid ?

6. Women's Records in Track and Field

Following, is a list of world records for women in running, from 100-meters to 10,000 meters.

Distance (m)	Time (s)	Avg Velocity (m/s)	Record Holder
100	10.49	9.53	Florence Joyner (USA) 1988
200	21.34	9.37	Florence Joyner (USA) 1988
400	47.60	8.40	Maria Koch (GER) 1985
800	113.28	7.06	Jarmila Kratochvilova (CZ) 1983
1000	149.34	6.69	Maria Mutola (MOZ) 1995
1500	230.46	6.51	Qu Yunxia (CHINA) 1993
1 mile	252.56	6.37	Svetlana Mastercova (RUS) 1996
2000	325.36	6.15	Sonia O'Sullivan (IRE) 1994
3000	486.11	6.17	Wang Junxia (CHINA) 1993
5000	864.53	5.78	Meseret Defar (ETH) 2006
10000	1771.78	5.64	Wang Junxia (CHINA) 1993

See if the Physics model of running will account for all the above world records. Try the following physiological constants

$$F = 1.2 * 9.8 \text{ m/s}^2, \quad k = 1.14 \text{ s}^{-1}, \quad \sigma = 36.5 \text{ W/kg}, \quad E_o = 2300 \text{ J/kg}$$

Based on the data fit, which of the above world records is most likely to be broken ?

7. Black Hole Sun

At the center of the Milky Way, at least a dozen stars orbit an unseen object. One of these stars, named S2, completes an orbit in 15 years and approaches within 17 lighthours of the center. The spectrum of S2 is that of a main-sequence star 15 times the mass of the sun.

If the maximum excursion of S2 from the center is 10 lightdays, then what is the mass of the unseen object ?

Chapter 4
Partial Differential Equations and Quantum Mechanics

4.1 Introduction to Partial Differential Equations

Some of the most important Physics equations are Partial Differential Equations, or PDEs. Let's examine one of the most famous, the Partial Differential Equation for Heat Diffusion, to see how an analytic solution may be obtained with standard mathematical techniques, and then obtain a computer solution of the same equation using *Mathematica.*

Partial Differential Equation #1

The partial differential equation for heat diffusion

$$\frac{\partial u}{\partial t} = \alpha \frac{\partial^2 u}{\partial x^2} \qquad \alpha = .96;$$

Boundary condition $u(x, t) = 0$ at $x = 0$ and $x = 20$

Initial condition $u(x, 0) = 70$ for $0 < x < 20$

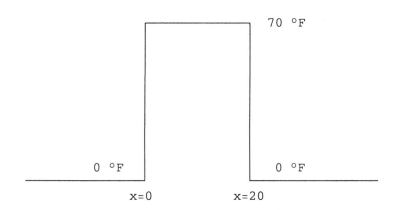

70 °F

0 °F   0 °F

x=0   x=20

Temperature distribution at $t = 0$

The Physics of this heat problem is very straightforward. Over time, the heat constrained between $x = 0$ and $x = 20$ will diffuse outward and be lost to the system, so the final steady-state distribution will be a flat zero across the board. The question is, how long will it take for the heat to leave the system, and what does the temperature distribution look like as a function of time?

Classical Solution by separation of variables

$$\frac{\partial u}{\partial t} = \alpha \frac{\partial^2 u}{\partial x^2}$$

Boundary condition   u (x, t) = 0 at x = 0 and x = 20
Initial condition   u (x, 0) = f[x] = 70 for 0 < x < 20

Let   u (x, t) = X (x) · T (t)

Then $\dfrac{\partial u}{\partial t}$ = X T′ = $\alpha \dfrac{\partial^2 u}{\partial x^2}$ = α X″ T

$$\frac{T'}{\alpha T} = \frac{X''}{X} = -\lambda^2$$

Thus   T (t) = A $e^{-\alpha \lambda^2 t}$ and   X (x) = B Sin λ x

Now match the boundary conditions at x = 0 and x = L,
where u (0, t) = u (L, t) = 0,

$$X = B_N \; Sin \; \frac{N \pi x}{L} \qquad N = 1, 2, 3 \ldots$$

and   X (x) = $\displaystyle\sum_{N=1}^{\infty} B_N \; Sin \; \frac{N \pi x}{L}$ note $\lambda = \dfrac{N \pi}{L}$

Then the solution is given by

$$u (x, t) = \sum_{N=1}^{\infty} A \, e^{-\alpha \lambda^2 t} \cdot B_N \; Sin \; \frac{N \pi x}{L} = \sum_{N=1}^{\infty} C_N \cdot e^{-\alpha \frac{N^2 \pi^2}{L^2} t} \cdot Sin \; \frac{N \pi x}{L}$$

We find the C_N by utilizing the Euler formula,

$$C_N = \frac{2}{L} \int_0^L f (x) \cdot Sin \; \frac{N \pi x}{L} \, dx \qquad \text{where } u (x, 0) = f (x) = 70$$

$$C_N = \frac{2}{L} \int_0^L 70 \cdot Sin \; \frac{N \pi x}{L} \, dx = \frac{2 * 70}{N \pi} (1 - Cos \, N \pi) = \frac{4 * 70}{N \pi} \quad (N \; odd)$$

$$C_N = \frac{4 * 70}{(2 k + 1) \pi} \qquad k = 0, 1, 2, 3 \ldots$$

For a grand solution of

$$u (x, t) = \frac{4 * 70}{\pi} \sum_{k=0}^{\infty} \frac{1}{(2 k + 1)} e^{-\alpha \frac{(2 k+1)^2 \pi^2}{L^2} t} \cdot Sin \; \frac{(2 k + 1) \pi x}{L}$$

Now, let us compare the analytic solution to the *Mathematica* solution. Using **NDSolve** we solve for **u[x,t]** and then 3-D plot the result

```
In[11]:=
  L = 20; α = .96;
  eq4 = {D[u[x, t], t] - .96*D[u[x, t], x, x] == 0,
    u[x, 0] == Which[x ≤ 0, 0, 0 < x < 20, 70, x ≥ 20, 0],
    u[0, t] == 0, u[20, t] == 0};
  sol4 = NDSolve[eq4, u[x, t], {x, 0, 20}, {t, 0, 40.0}];
  Plot3D[Evaluate[u[x, t] /. sol4[[1]]], {t, 0, 40.0}, {x, 0, 20}],
    PlotRange → All, AxesLabel → {"t", "x", " "}];
```

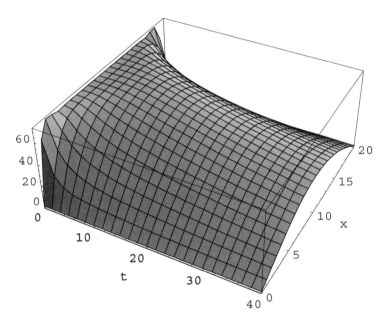

3-D representation of temperature in the slab versus time.

The *Mathematica* solution is retained in memory, so let us compare the *exact* solution to the numerical solution, by comparing the temperature at $x = 10$ (down the center-line) as a function of time.

$$\text{Table}\Big[$$

$$\text{Evaluate}\Big[\{t, u[x, t], \frac{4*70}{\pi} \sum_{k=0}^{10} \frac{(-1)^k}{(2k+1)} * \text{Exp}\Big[-\frac{\alpha(2k+1)^2}{(L/\pi)^2} t\Big]\}\Big] /.$$

$$\text{sol4}[[1]] /. x \to 10, \{t, 2.0, 40.0, 2.0\}\Big]\Big] // \text{TableForm}$$

| | *Mathematica* | Exact |
t (da)	T (x = 10)	T (x = 10)
4.	69.9683	69.9569
8.	68.4981	68.4986
12.	64.7889	64.7891
16.	60.0307	60.0324
20.	55.0783	55.0785
24.	50.3008	50.3015
28.	45.8394	45.8404
32.	41.7318	41.7333
36.	37.9752	37.9763
40.	34.549	34.55

The agreement is very very good.

Note that we may also plot just temperature versus time as a 2-D graphic, and, as expected, the temperature of the system decreases over time as heat leaves the system.

```
In[47]:=  f[x_ , t_] = u[x, t] /. sol4[[1]];
          Pictures = Table[f[x, t], {t, 0.5, 30.5, 3}]
          Plot[Evaluate[Pictures], {x, 0, 20}]
```

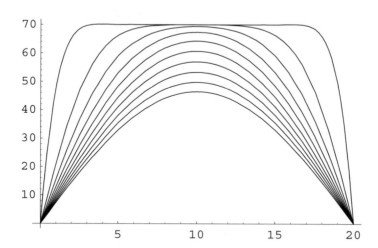

The temperature distribution with time. Notice how the originally rectangular temperature distribution becomes more and more like a sine function as time goes on.

Note that it is not whether the *Mathematica* or the exact solution is better, it is that the *Mathematica* computer solution is easier to use.

Partial Differential Equation #2

$$\frac{\partial u}{\partial t} = \alpha \frac{\partial^2 u}{\partial x^2} \qquad \text{Let } \alpha = 1, \quad L = 1$$

Boundary condition $u(x, t) = 0$ at $x = 0$ and $x = 1$

Initial condition $\quad u(x, 0) = f(x) = 100 * \text{Sin}(\pi x)$ for $0 < x < 1$

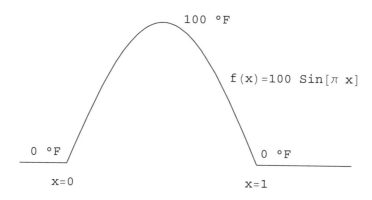

100 °F

$f(x) = 100 \, \text{Sin}[\pi \, x]$

0 °F   0 °F

x=0   x=1

Temperature distribution at $t = 0$

Classical Solution by separation of variables

Let $\quad u(x, t) = X(x) \cdot T(t)$

Then $\quad \dfrac{\partial u}{\partial t} = X T' = \alpha \dfrac{\partial^2 u}{\partial x^2} = \alpha X'' T$

$$\frac{T'}{\alpha T} = \frac{X''}{X} = -\lambda^2$$

Thus $\quad T(t) = A e^{-\alpha \lambda^2 t} \qquad$ and $\quad X(x) = B \, \text{Sin} \, \lambda x$

Now match the boundary conditions at $x = 0$ and $x = L$, where $u(0, t) = u(L, t) = 0$,

$$X = B_N \, \text{Sin} \, \frac{N \pi x}{L} \qquad N = 1, 2, 3 \ldots$$

and $\quad X(x) = \displaystyle\sum_{N=1}^{\infty} B_N \, \text{Sin} \, \frac{N \pi x}{L} \qquad$ note $\lambda = \dfrac{N \pi}{L}$

Then the solution is given by

$$u(x, t) = \sum_{N=1}^{\infty} A e^{-\alpha \lambda^2 t} \cdot B_N \, \text{Sin} \, \frac{N \pi x}{L} = \sum_{N=1}^{\infty} C_N \cdot e^{-\alpha \frac{N^2 \pi^2}{L^2} t} \cdot \text{Sin} \, \frac{N \pi x}{L}$$

We find the C_N by utilizing the Euler formula,

$$C_N = \frac{2}{L} \int_0^L f(x) \cdot \text{Sin} \frac{N \pi x}{L} dx \quad \text{where } u(x, 0) = f(x) = 100 \, \text{Sin}[\pi x]$$

$$C_N = \frac{2}{1} \int_0^1 100 \, \text{Sin}[\pi x] \cdot \text{Sin}[N \pi x] \, dx$$

By *orthogonality* of the sine functions
$$C_N = 100 \quad \text{if } N = 1$$
$$C_N = 0 \quad \text{if } N = 2, 3, 4 \ldots$$

The exact solution is
$$u(x, t) = 100 \, e^{-\pi^2 t} \cdot \text{Sin} \, \pi x \quad \text{for } 0 \le x \le 1$$

Now, let us compare the analytic solution to the *Mathematica* solution, using **NDSolve**

```
In[4]:=  eq2 = {D[u[x, t], t] - D[u[x, t], x, x] == 0,
           u[x, 0] == 100 Sin[π*x], u[0, t] == 0, u[1, t] == 0};
         sol2 = NDSolve[eq2, u[x, t], {x, 0, 1}, {t, 0, 0.5}];
         Plot3D[Evaluate[u[x, t] /. sol2[[1]]], {t, 0, 0.5}, {x, 0, 1}],
           PlotRange → All, AxesLabel → {"t", "x", " "}];
         Table[Evaluate[{t, u[x, t], 100 * E ^ (-π^2 * t)} /. sol2[[1]] /.
           x → 0.5, {t, 0, .5, .1}]] // TableForm
```

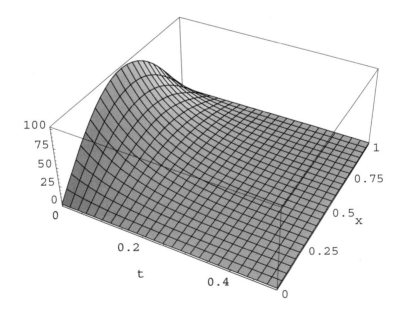

3-D representation of decrease of sine-temperature function with time

Comparing the *Mathematica* **NDSolve** and the exact solutions at x = 0.5,

	Mathematica	Exact Solution
t	T (x = 0.5)	T (x = 0.5)
0	100.	100
0.1	37.2729	37.2708
0.2	13.8913	13.8911
0.3	5.17645	5.17733
0.4	1.9287	1.92963
0.5	0.718935	0.719188

Again, the numbers are very very close.

We will also show the temperature change in the system with time, as a 2-D representation

```
In[86]:=  f[x_, t_] = u[x, t] /. sol2[[1]];
          Pictures = Table[f[x, t], {t, 0, 0.3, .05}]
          Plot[Evaluate[Pictures], {x, 0, 1},
            Epilog → {Text["t=0", {0.72, 88}], Text["t=.05", {.67, 61}],
              Text["t=.10", {.62, 41}], Text["t=.15", {.58, 26}]}]
```

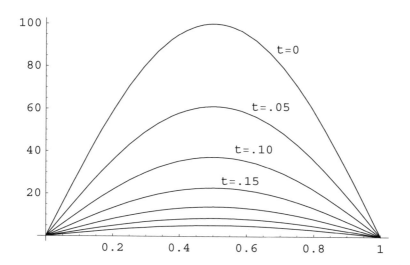

Temperature decrease with time, as heat diffuses out of the system.

With the numbers between *Mathematica* and the exact solutions so close, we will now proceed to analyze even more complicated heat problems using **NDSolve**. We will compare the *exact* solution, whenever available, with the *Mathematica* solution of the Partial Differential Equation.

4.2 PDE – Copper Rod

One end of a 40-cm long copper rod of cross-section 10 cm² is buried in ice at 0 °C, and the free end is heated with a propane torch to 400 °C. How long before the temperature gradient is *linear* (within 1%) from 400 °C to 0 °C along the length of the rod? Assuming heat-transfer by conduction only (if we insulate the rod to reduce convection and radiation losses), how much heat energy per second is transmitted by the copper rod from the flame to the ice?

The thermal conductivity of copper is 390 W/m/°C. The thermal diffusivity of copper is $\alpha = 0.0061\,\text{m}^2/\text{min}$. The PDE we are solving is

$$\frac{\partial u}{\partial t} = \alpha \frac{\partial^2 u}{\partial x^2} \qquad \alpha = .0061\,\text{m}^2/\text{min}$$

Boundary condition $u(x, t) = 0$ at $x = 0$ and $u(x, t) = 400$ at $x = 0.4\,\text{m}$
Initial condition $u(x, 0) = 0$ for $0 \leq x < 0.4\,\text{m}$

Note that we have what are apparently two contradictory conditions: we want the initial temperature of the rod to be 0 °C all along its length at $t = 0$, *and* we want the free end to be heated to 400 °C. How do we reconcile these two conditions?

The answer is that we allow the free end to heat up rapidly with a temperature function $u[.4,t]= 400-400*\text{Exp}[-1000t]$. Now both the boundary conditions and the initial condition are satisfied, and we also have the physical effect of rapidly heating the end of the rod with the propane torch.

The *Mathematica* programming to find $u[x,t]$ is straightforward

```
eq4 = {D[u[x, t], t] - .0061 * D[u[x, t], x, x] == 0,
    u[x, 0] == 0, u[0, t] == 0, u[.4, t] == 400 - 400 * Exp[-1000 t]};
sol400 = NDSolve[eq4, u[x, t], {x, 0, .4}, {t, 0, 20}];
Clear@f;  f[x_, t_] = u[x, t] /. sol400[[1]]
```

However the graphing of $u[x,t]$ is more complex

```
Show[
  Plot3D[Evaluate[u[x, t] /. sol400[[1]], {t, 0, 10}, {x, 0, .4}],
    PlotRange → All, AxesLabel → {"t", "x", " "},
    Ticks →
      {Table[{2 * k, 2 * (5 - k)}, {k, 0, 5, 1}], Automatic, Automatic},
    DisplayFunction → Identity] /. SurfaceGraphics[arr_, opts___] :>
    SurfaceGraphics[Reverse /@ arr, opts],
  DisplayFunction → $DisplayFunction,
  ViewPoint -> {2.428, -2.280, 1.010}]
Pictures = Table[f[x, t], {t, 0, 11}]
Plot[Evaluate[Pictures], {x, 0, 0.4}, Frame → True,
  Prolog → {Text["t=0 min", {0.35, 10}], Text["1 min" , {.30, 85}],
    Text["2", {.259, 123}], Text["3", {.241, 146}]}]
```

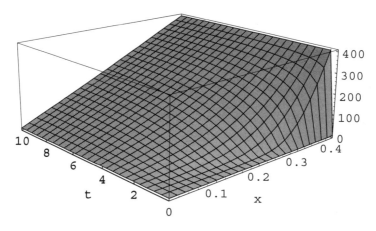

3-D representation of temperature in the copper rod

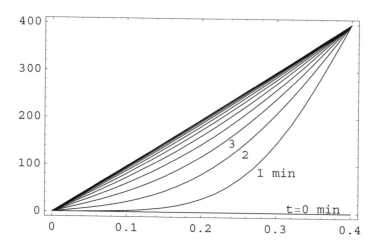

2-D plot of temperature vs distance along the rod, from t = 0 to t = 11 min

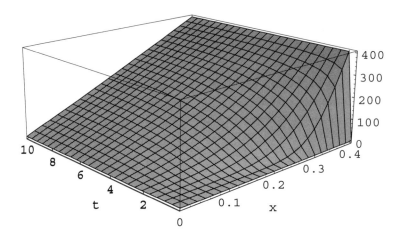

If we **Table** the temperature values, we will find that the temperature gradient becomes linear (within 1%) somewhere between t = 10 min and t = 14 min.

```
TableForm[Transpose@Prepend[
   Transpose@Table[f[x, t], {t, 0, 15, 1}, {x, 0, .40, 0.10}],
   Table[StyleForm[ToString@t, FontWeight → "Bold"],
    {t, 0, 15, 1}]],
  TableHeadings -> {None, Prepend[Table[StyleForm["x(" <>
      ToString@x <> ")", FontWeight → "Bold"], {x, 0, .40, 0.10}],
    StyleForm["t(min)", FontWeight → "Bold"]]}] // Chop
```

t(min)	x(0)	x(0.1)	x(0.2)	x(0.3)	x(0.4)
0	0	0	0	0	0
1	0	2.62697	28.0163	146.007	399.968
2	0	21.3455	80.0686	208.763	399.968
3	0	43.1353	117.609	240.328	399.968
4	0	60.3177	143.438	259.682	399.968
5	0	72.6146	161.169	272.462	399.968
6	0	81.165	173.335	281.115	399.968
7	0	87.0546	181.68	287.028	399.968
8	0	91.103	187.411	291.084	399.968
9	0	93.8899	191.354	293.874	399.968
10	0	95.8083	194.068	295.793	399.968
11	0	97.1227	195.926	297.107	399.968
12	0	98.023	197.199	298.007	399.968
13	0	98.6414	198.074	298.625	399.968
14	0	99.0665	198.675	299.051	399.968
15	0	99.3581	199.088	299.342	399.968

Note that when the temperature gradient is approximately linear that the straight-line equation of heat conduction applies:

$$\dot{Q} = \frac{k\,A\,\Delta T}{\Delta x} = \frac{390 * .001 * 400}{0.4} = 390 \text{ Watts}$$

4.3 Heat Conduction in the Earth

A Temperature Wave into the Earth

A classic problem in mechanical engineering (Incropera and DeWitt, *Heat Transfer*) is to determine how deep water pipes should be buried such that they will never freeze.

If we assume that in most northern latitudes of the United States (Alaska excepted) that the temperature at the Earth's surface varies from + 35 °C in summer to −15 °C in winter then T (surface) = 10 + 25 Sin [2π t] where t is measured in years

Let us assume that the Partial Differential Equation for heat transfer in soil is

$$\frac{\partial u}{\partial t} = 2.4 \; \frac{\partial^2 u}{\partial x^2} \qquad \text{where} \quad \alpha = 2.4 \, \text{m}^2 / \text{yr}$$

with initial conditions u (x, 0) = 10 [the ground temperature **u** = 10 at all depths at t = 0],
the surface temperature at any time t is u (0, t) = 10 + 25 Sin [2π t], and
an additional boundary condition (D[u[x, t], x] /. x → 4) = 0 must also be imposed

Solving with *Mathematica* we find a highly attenuated temperature wave propagating into the Earth.

```
In[12]:=   eq3 = {D[u[x, t], t] - 2.40 * D[u[x, t], x, x] == 0, u[x, 0] == 10,
             u[0, t] == (10 + 25 * Sin[2 π * t]), (D[u[x, t], x] /. x → 4) == 0};
           sol1 = NDSolve[eq3, u[x, t], {x, 0, 4}, {t, 0, 3}];
           Plot3D[Evaluate[u[x, t] /. sol1[[1]], {x, 0, 2}, {t, 0, 3}]];
```

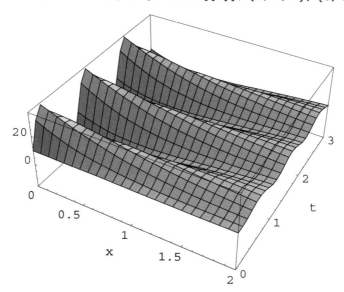

Shown above is the temperature variation at the Earth's surface on the left and how it propagates into the earth with x in meters. The temperature variation is phase-shifted as one goes deeper into the Earth, because it takes a certain length of time before the heat diffuses to a given depth.

How deep, then, should we bury the water pipes? A table of values will determine the answer. At whatever depth the temperature never goes negative is the depth at which the water pipes wont freeze. From the table, any depth greater than x = 0.8 meters is safe from a –15 °C surface freeze in winter.

```
Clear@f; f[x_, t_] = u[x, t] /. sol1[[1]]; TableForm[Transpose@
  Prepend[Transpose@Table[f[x, t], {t, 0, 2.0, .1}, {x, 0, 1.0, 0.2}],
    Table[StyleForm[ToString@t, FontWeight → "Bold"], {t, 0, 2.0, .1}]],
  TableHeadings -> {None, Prepend[
    Table[StyleForm["x(" <> ToString@x <> ")", FontWeight → "Bold"],
    {x, 0, 1.0, 0.2}], StyleForm["t(yr)", FontWeight → "Bold"]]}]
```

t(yr)	**x(0)**	**x(0.2)**	**x(0.4)**	**x(0.6)**	**x(0.8)**
0	10.	10.	10.	10.	10.
0.1	24.694	19.188	15.460	13.070	11.631
0.2	33.776	27.877	22.987	19.122	16.201
0.3	33.777	30.392	26.782	23.332	20.265
0.4	24.696	25.462	24.828	23.373	21.518
0.5	10.002	14.846	17.637	18.911	19.124
0.6	-4.692	2.5371	7.8350	11.487	13.798
0.7	-13.77	-6.801	-0.905	3.8337	7.4515
0.8	-13.77	-9.627	-5.294	-1.192	2.4257
0.9	-4.693	-4.878	-3.687	-1.719	0.5822
1.	10.000	5.6206	3.2783	2.4217	2.5836
1.1	24.695	17.849	12.924	9.6239	7.6337
1.2	33.778	27.132	21.554	17.118	13.781
1.3	33.779	29.917	25.862	22.027	18.658
1.4	24.704	25.140	24.195	22.465	20.389
1.5	10.010	14.617	17.184	18.257	18.302
1.6	-4.684	2.3697	7.5001	11.001	13.184
1.7	-13.76	-6.927	-1.159	3.4640	6.9814
1.8	-13.76	-9.723	-5.490	-1.480	2.0578
1.9	-4.685	-4.952	-3.841	-1.947	**0.2880**
2.	10.009	5.5615	3.1540	2.2361	2.3429

It is worthwhile to check our answer against the **exact** analytic solution which is

$$T(x,t) = 10 + 25 \exp(- x \sqrt{\pi f / \alpha})* \sin[2\pi f t - x \sqrt{\pi f / \alpha}]$$

Then, the maximum depth at which freezing will occur is found by solving

$In[2]:=$ $\mathtt{Solve\left[0 == 10 + 25 * Exp\left[-x \sqrt{\pi * 1/2.4}\right] * (-1), x\right]}$

$Out[2]=$ $\{\{x \to 0.800874\}\}$

4.4   Quantum Physics

One of the most puzzling and also the most interesting aspects of twentieth-century Physics is the advent of Quantum mechanics. How is it possible that an electron or an atom or a nucleus may be <u>both</u> a particle and a wave? The answer is *that is just how Nature is*. The value of Quantum Physics is that it accurately describes Nature at the atomic level. Given a particular physical situation, Quantum theory will give the correct energy *eigenstates* of the system and the *probability* of a process occuring.

The fundamental equation of Quantum Physics is Schrodinger's equation

$$\frac{-\hbar^2}{2\,m}\,\psi'' + V\,\psi = E\,\psi$$

We shall only use the one-dimensional case in this and the following section, because the mathematical difficulties rapidly become formidable. For a very instructive discussion on Schrodinger's equation, see *"Feynman's Derivation of Schrodinger's Equation"* by David Derbes in the July 1996 American Journal of Physics.

The Schrodinger equation may also be written as

$$\psi'' = -\frac{2\,m}{\hbar^2}\,(E - V)\,\psi$$

or more succinctly as

$$\psi'' = -k^2\psi \qquad \text{where} \qquad k = \frac{\sqrt{2\,m\,(E-V)}}{\hbar}$$

where the wavefunction solutions are of the form

$$\psi = A\,e^{i\,k\,x} \qquad \text{where A, k, and } \psi \text{ may be } \underline{\text{complex}}$$

In the following sections, we show how *Mathematica* may be used in quantum computations to match wavefunctions and their derivatives -- and find the probability of a process occuring. We will consider three examples: the Quantum Step, the Quantum Barrier, and the Quantum Well. Then we will utilize the *Mathematica* program with the *Shooting Method* to find the energy eigenstates and the wavefunctions for the Quantum Oscillator and the Hydrogen Atom.

4.4 A Quantum Step

A 7-eV electron encounters a 5-eV potential step. What is the probability that the electron is transmitted? What is the probability that it is reflected?

```
In[9]:= Plot[{7, 5 * UnitStep[x - 1]}, {x, -5, 5}, PlotRange → {0, 10.1},
        Prolog → {Text["7 eV", {4.2, 7.5}],
          Text["5 eV", {4.6, 4.2}], Text["Region 1", {-3.4, 1.5}],
          Text["Region 2", {2.8, 1.5}]}]
```

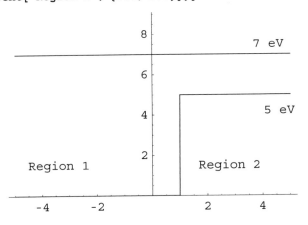

Potential Step

Classically, we would expect every electron encountering the potential step to be transmitted, albeit at a lower kinetic energy. However, Quantum Physics says that there is a probability that the electron will be reflected.

We are looking for solutions of the form

$$\psi_1 = e^{ikx} + r e^{-ikx} \qquad \text{where} \quad k = \frac{\sqrt{2\,m\,E}}{\hbar}$$

$$\psi_2 = t e^{i\alpha x} \qquad \text{where} \quad \alpha = \frac{\sqrt{2\,m\,(E-V)}}{\hbar}$$

$$k = \sqrt{2 * 9.1 * 10^{-31} * 7 * 1.6 * 10^{-19}} \Big/ (6.626 * 10^{-34} / (2\,\pi)) \Big/ 10^{10}$$

$$\alpha = \sqrt{2 * 9.1 * 10^{-31} * 2 * 1.6 * 10^{-19}} \Big/ (6.626 * 10^{-34} / (2\,\pi)) \Big/ 10^{10}$$

```
Out[10]= 1.35386
```

```
Out[11]= 0.723668
```

```
In[10]:= Clear[r, t]; k = 1.3538; α = .7236;
         sol = NSolve[{1 + r == t, k * (1 - r) == α * t}, {r, t}]
         R = Abs[r]^2 /. sol
         T = (α / k) * Abs[t]^2 /. sol
         psi = Re[Exp[I * k * x] + r * Exp[-I * k * x]] /. sol;
         bar = Re[t * Exp[I * α * x]] /. sol;
         Plot[psi * If[x < 0, 1, 0] + bar * (UnitStep[x]), {x, -3.5, 13}]
```

Out[12]= {{r → 0.30336, t → 1.30336}}

Out[13]= {R=0.0920273}

Out[14]= {T=0.907973}

We find, using *Mathematica*, that 9.2% of the incident electron beam is reflected, and 90.8% is transmitted.

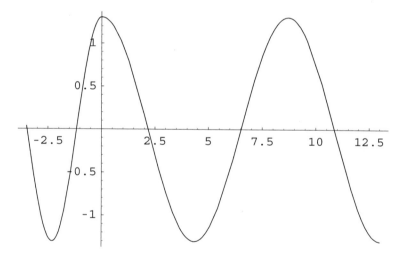

The real part of the wavefunction

The wavefunction of the transmitted wave is shown proceeding to the right with longer wavelength (less kinetic energy). The wavefunction to the left consists of both the incident and reflected wave.

4.4 B   Quantum Barrier

A 3-eV electron encounters a 5-eV potential barrier. The barrier is 2 Angstroms wide. What is the probability the electron tunnels through ?

```
In[219]:=  Plot[{3, 5 * UnitStep[x - 1.1] - 5 * UnitStep[x - 3.1]},
             {x, -3, 4}, PlotRange → {0, 8.1},
             Prolog → {Text["3 eV", {3.8, 3.4}], Text["5 eV", {3.6, 5}]}]
```

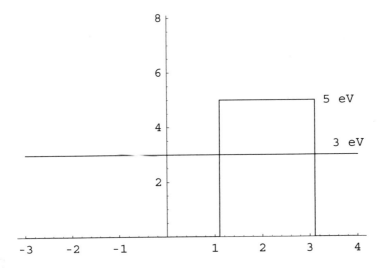

The Quantum wavefunctions in the three regions are :

$$\psi_1 = e^{ikx} + r e^{-ikx} \qquad \text{where} \quad k = \frac{\sqrt{2 \, m \, E}}{\hbar}$$

$$\psi_2 = a \, e^{\alpha x} + b \, e^{-\alpha x} \qquad \text{where} \quad \alpha = \frac{\sqrt{2 \, m \, (V - E)}}{\hbar}$$

$$\psi_3 = t \, e^{ikx}$$

$$k = \sqrt{2 * 9.1 * 10^{-31} * 3 * 1.6 * 10^{-19}} \Big/ (6.626 * 10^{-34} / (2 \, \pi)) \Big/ 10^{10}$$

$$\alpha = \sqrt{2 * 9.1 * 10^{-31} * 2 * 1.6 * 10^{-19}} \Big/ (6.626 * 10^{-34} / (2 \, \pi)) \Big/ 10^{10}$$

```
Out[63]=  0.886308
```

```
Out[64]=  0.723668
```

```
In[204]:= Clear[a, b, r, t];
          k = .886; α = .724; L = 2;
          sol = NSolve[
            {1 + r == a + b, a * Exp[α * L] + b * Exp[-α * L] == t * Exp[I * k * L],
             I * k - I * k * r == α * a - α * b, α * a * Exp[α * L] - α * b * Exp[-α * L] ==
               I * k * t * Exp[I * k * L]}, {a, b, r, t}]
          R = Abs[r]^2 /. sol
          T = Abs[t]^2 /. sol
          psi = Re[Exp[I * k * x] + r * Exp[-I * k * x]] /. sol;
          bar = Re[a * Exp[α * x] + b * Exp[-α * x]] /. sol;
          tran = Re[t * Exp[I * k * x]] /. sol;
          Plot[
           psi * If[x < 0, 1, 0] + bar * (UnitStep[x] - UnitStep[x - 2]) +
            tran * If[x > 2, 1, 0], {x, -10, 10}]
```

Out[206]= {{a → 0.039386 + 0.0712345 i, b → 1.12157 - 0.955549 i,
              r → 0.160958 - 0.884315 i, t → -0.00927323 - 0.438172 i}}

Out[207]= {R=0.80792}

Out[208]= {T=0.19208}

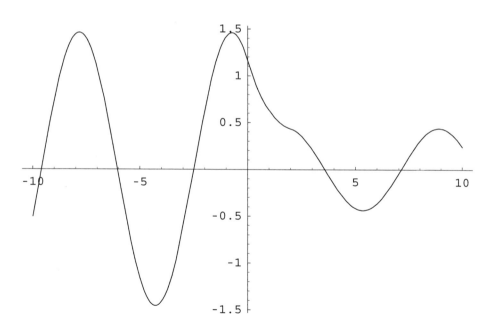

The real part of the wavefunction. 19% of the incident electron beam
is transmitted. Classically, we would expect <u>none</u> of the electron beam to
be transmitted.

4.4 C Quantum Well

A 3-eV electron encounters an 8-eV potential well. The well is 3 Angstroms wide. What is the probability of transmission? What is the probability the electron is reflected?

```
Plot[{3, -8 * UnitStep[x - 1.1] + 8 * UnitStep[x - 3.1]},
  {x, -3, 4}, PlotRange → {-8.3, 4.1}, Axes → None,
  Prolog → {Text["3 eV", {3.8, 3.4}], Text["-8 eV", {3.6, -7.5}]}]
```

The Quantum wavefunctions in the three regions are :

$$\psi_1 = e^{ikx} + r e^{-ikx} \qquad \text{where } k = \frac{\sqrt{2\, m\, E}}{\hbar}$$

$$\psi_2 = a\, e^{i\alpha x} + b\, e^{-i\alpha x} \qquad \text{where } \alpha = \frac{\sqrt{2\, m\, (E + |V|)}}{\hbar}$$

$$\psi_3 = t\, e^{ikx}$$

$In[16]:=$ $k = \sqrt{2 * 9.1 * 10^{-31} * 3 * 1.6 * 10^{-19}} \Big/ (6.626 * 10^{-34} / (2\pi)) \Big/ 10^{10}$

$\alpha = \sqrt{2 * 9.1 * 10^{-31} * 11 * 1.6 * 10^{-19}} \Big/ (6.626 * 10^{-34} / (2\pi)) \Big/ 10^{10}$

$Out[16]=$ 0.886308

$Out[17]=$ 1.69715

We may use *Mathematica* to match the wavefunctions and their derivatives at $x = 0$ and $x = L$

```
Clear[a, b, r, t];
 k = .886; α = 1.697; L = 3;
sol = NSolve[
   {1 + r == a + b, a * Exp[I * α * L] + b * Exp[-I * α * L] == t * Exp[I * k * L],
    k - k * r == α * a - α * b, α * a * Exp[I * α * L] - α * b * Exp[-I * α * L] ==
    k * t * Exp[I * k * L]}, {a, b, r, t}]
R = Abs[r]^2 /. sol
T = Abs[t]^2 /. sol
psi = Re[Exp[I * k * x] + r * Exp[-I * k * x]] /. sol;
bar = Re[a * Exp[I * α * x] + b * Exp[-I * α * x]] /. sol;
tran = Re[t * Exp[I * k * x]] /. sol;
Plot[psi * If[x < 0, 1, 0] + bar * (UnitStep[x] - UnitStep[x - 3]) +
   tran * If[x > 3, 1, 0], {x, -10, 10}]
```

Out[192]=

$\quad$ {{a → 0.637615 - 0.0402901 i, b → -0.154181 - 0.128322 i,
$\qquad$ r → -0.516566 - 0.168612 i, t → -0.60168 + 0.585415 i}}

Out[193]= {R = 0.295271}

Out[194]= {T = 0.704729}

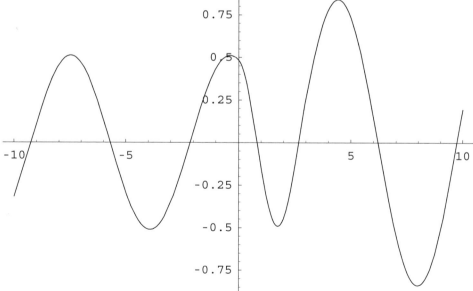

The real part of the wavefunction. 70% of the incident electron beam is transmitted. Classically, we would expect <u>all</u> of the electron beam to be transmitted.

4.4 D Quantum Oscillator

A particle of mass m is inside a potential well of depth $V = \frac{1}{2} kx^2$. How do we find the allowed quantum energy levels?

```
Plot [x² / 2, {x, -5, 5}, Prolog → {Text ["V= 1/2 kx²", {3.4, 11.4}],

   {GrayLevel[0.5], Line[{{-2.8, 1}, {2.7, 1}}],
   Line[{{-3.68, 3}, {3.56, 3}}], Line[{{-4.62, 5}, {4.4, 5}}]}}]
```

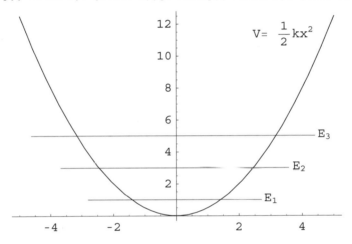

Discrete energy levels in a harmonic oscillator potential

Let's apply the one-dimensional Schrodinger equation to find the allowed energy levels, E_1, E_2, E_3, ...

$$\frac{-\hbar^2}{2\,m} \psi'' + V\,\psi = E\,\psi$$

for a particle of mass m in a harmonic-oscillator potential $V(x) = \frac{1}{2} kx^2$

$$\psi'' = -\frac{2\,m}{\hbar^2}(E - \tfrac{1}{2} k\,x^2)\,\psi$$

$$\psi'' = -\frac{2\,m}{\hbar^2}(E - m\,\omega^2 \frac{x^2}{2})\,\psi \qquad \text{setting } \omega = \sqrt{k/m}$$

$$\psi'' = -(\frac{2\,m}{\hbar^2} E - \frac{m^2\,\omega^2}{\hbar^2} x^2)\,\psi$$

Let $\beta = \frac{2\,m}{\hbar^2} E$ and $\alpha = \frac{m\,\omega}{\hbar}$

$$\psi'' = -(\beta - \alpha^2 x^2)\,\psi$$

Then if $\epsilon = \frac{\beta}{\alpha}$ and $z = \sqrt{\alpha}\,x$,

$$\frac{d^2\psi}{dz^2} = -(\epsilon - z^2)\,\psi$$

This is the equation we wish to solve using *Mathematica*.

Notice that we have chosen $\epsilon = \frac{\beta}{\alpha} = \frac{2E}{\hbar\omega} = \frac{2E}{h\nu}$ where ν is the frequency of oscillation. So once we determine the *eigenvalues* for ϵ, then we have the allowable energies for the particle in the well. That is, there are only certain values of ϵ that will allow ψ to go to zero as $z \to \pm\infty$.

Note also that in the harmonic oscillator equation, $\dfrac{d^2\psi}{dz^2} = -(\epsilon - z^2)\psi$ that as $z \to 0$ then $\psi'' = -\epsilon\psi$ with solutions of cosines and sines. Therefore, for starting values of ψ choose $\psi(0) = 1$ and $\psi'(0) = 0$ <u>or</u> $\psi(0) = 0$ and $\psi'(0) = 1$. (We will normalize ψ later).

<u>The Shooting Method for solving Schrodinger's Equation</u>

The strategy we shall follow in finding the energy levels and acceptable wavefunctions is as follows:

We shall start with some initial value of ϵ with $\psi(0) = 1$ and $\psi'(0) = 0$ and then observe whether the wavefunction goes to zero as $z \to \infty$. This is the criterion for a particle to be bound in a quantum well.

As an example, let's start with $\epsilon = 1$

```
In[11]:= Clear[y, z]; ε = 1;
         sol1 =
         DSolve[{y''[z] == -(ε - z^2) y[z], y[0] == 1, y'[0] == 0}, y[z], z]
         Plot[Evaluate[y[z] /. sol1, {z, -5, 5}]];
```

$$Out[12] = \left\{\left\{y[z] \to e^{-\frac{z^2}{2}}\right\}\right\}$$

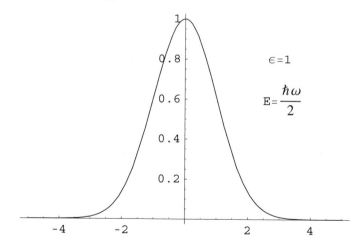

An acceptable wavefunction is found for $\epsilon = 1$

We now normalize the wavefunction using the substitution $z = \sqrt{\alpha}\, x$

$In[7]:=$ **Assuming$\left[\text{Re}[\alpha] > 0,\ \text{Solve}\left[A^2 \int_{-\infty}^{\infty} \text{Exp}[-\alpha * x^2]\ dx == 1,\ A\right]\right]$**

$Out[7]=$ $\left\{\left\{A \to -\dfrac{\alpha^{1/4}}{\pi^{1/4}}\right\},\ \left\{A \to \dfrac{\alpha^{1/4}}{\pi^{1/4}}\right\}\right\}$

Therefore, $\psi_1(x) = \dfrac{\alpha^{1/4}}{\pi^{1/4}} e^{-\alpha x^2/2}$ and $E_1 = 1\ \dfrac{\hbar\omega}{2}$

The next eigenvalue occurs at $\epsilon = 3$,

$In[1]:=$ **Clear[y, z]; $\epsilon = 3$;**
sol2 =
DSolve[{y''[z] == -(3 - z^2) y[z], y[0] == 0, y'[0] == 1}, y[z], z]
Plot$\left[\text{Evaluate}[y[z]\ /.\ \text{sol2}, \{z, -5, 5\}],\right.$
$\left.\text{Prolog} \to \left\{\text{Text}["\epsilon=3", \{3, .58\}],\ \text{Text}\left["E=\dfrac{3\hbar\omega}{2}",\ \{3.1, .4\}\right]\right\}\right];$

$Out[2]=$ $\left\{\left\{y[z] \to e^{-\frac{z^2}{2}} z\right\}\right\}$

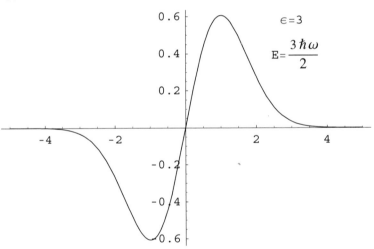

The wavefunction for $\epsilon = 3$.

$In[9]:=$ **Assuming$\left[\text{Re}[\alpha] > 0,\ \text{Solve}\left[A^2 \int_{-\infty}^{\infty} \alpha * x^2 * \text{Exp}[-\alpha * x^2]\ dx == 1,\ A\right]\right]$**

$Out[9]=$ $\left\{\left\{A \to -\dfrac{\sqrt{2}\ \alpha^{1/4}}{\pi^{1/4}}\right\},\ \left\{A \to \dfrac{\sqrt{2}\ \alpha^{1/4}}{\pi^{1/4}}\right\}\right\}$

Therefore, $\psi_2(x) = \sqrt{2}\ \dfrac{\alpha^{1/4}}{\pi^{1/4}} \sqrt{\alpha}\ x\ e^{-\alpha x^2/2}$ $E_2 = 3\ \dfrac{\hbar\omega}{2}$

The next eigenvalue occurs at $\epsilon = 5$,

```
ε = 5;
sol3 = DSolve[{y"[z] == - (5 - z^2) y[z], y[0] == 1, y'[0] == 0}, y[z], z]
Plot[Evaluate[y[z] /. sol3, {z, -5, 5}]];
```

$$Out[5] = \left\{\left\{ y[z] \rightarrow - e^{-\frac{z^2}{2}} (-1 + 2 z^2) \right\}\right\}$$

$\epsilon = 5$

$$E = \frac{5\, \hbar\omega}{2}$$

The wavefunction for $\epsilon = 5$.

$In[8]:=$ $\mathbf{Assuming}\left[\mathbf{Re[\alpha] > 0,}\right.$

$\qquad \mathbf{Solve}\left[A^2 \int_{-\infty}^{\infty} \mathbf{Exp[-\alpha * x^2] * (1 - 2\,\alpha * x^2)^2\, dx} == 1, A\right]\left.\right]$

$$Out[8] = \left\{\left\{ A \rightarrow -\frac{\alpha^{1/4}}{\sqrt{2}\ \pi^{1/4}} \right\}, \left\{ A \rightarrow \frac{\alpha^{1/4}}{\sqrt{2}\ \pi^{1/4}} \right\}\right\}$$

$$\psi_3(x) = \frac{\alpha^{1/4}}{\sqrt{2}\ \pi^{1/4}} (1 - 2\alpha x^2)\, e^{-\alpha x^2/2} \qquad E_3 = 5\, \frac{\hbar\omega}{2}$$

Therefore the wavefunctions for the first three energy levels are

$$\psi_1(x) = \frac{\alpha^{1/4}}{\pi^{1/4}} e^{-\alpha x^2/2} \qquad\qquad E_1 = 1\, \frac{\hbar\omega}{2}$$

$$\psi_2(x) = \sqrt{2}\, \frac{\alpha^{1/4}}{\pi^{1/4}} \sqrt{\alpha}\ x\, e^{-\alpha x^2/2} \qquad\qquad E_2 = 3\, \frac{\hbar\omega}{2}$$

$$\psi_3(x) = \frac{\alpha^{1/4}}{(4\,\pi)^{1/4}} (1 - 2\alpha x^2)\, e^{-\alpha x^2/2} \qquad\qquad E_3 = 5\, \frac{\hbar\omega}{2}$$

Notice that once the energy eigenvalue ϵ is found that the wavefunctions are easy to obtain. However, how does one find the *eigenvalues*?

The easiest way is just to examine a graph, as one gradually increases the value of ϵ. Following are the wavefunctions for $\epsilon = 0.9$ and $\epsilon = 1.1$. They both cascade off to infinity. But there is a change in direction in going from 0.9 to 1.1. Therefore ϵ is between 0.9 and 1.1

```
ε = 0.9; DSolve[{y"[z] == - (ε - z^2) y[z], y[0] == 1, y'[0] == 0}, y[z], z]
Plot[Evaluate[y[z] /. %, {z, -5, 5}]];
```

$Out[24] = \left\{\left\{y[z] \to e^{-0.5 z^2} \left(\text{Hypergeometric1F1}\left[0.025, \frac{1}{2}, 1. z^2\right]\right)\right\}\right\}$

unacceptable wavefunction ψ **goes to** $+\infty$ **for** z > 4 (ϵ = 0.9 **is too small**)

```
ε = 1.1; DSolve[{y"[z] == - (ε - z^2) y[z], y[0] == 1, y'[0] == 0}, y[z], z]
Plot[Evaluate[y[z] /. %, {z, -5, 5}]];
```

$Out[36] = \left\{\left\{y[z] \to e^{-0.5 z^2} \left(\text{Hypergeometric1F1}\left[-0.025, \frac{1}{2}, 1. z^2\right]\right)\right\}\right\}$

unacceptable wavefunction ψ **goes to** $-\infty$ **for** z > 4 (ϵ = 1.1 **is too large**)

Therefore the "shooting method" allows one to rapidly home-in on the eigenvalues. In the next section, we will show how the shooting method may be used with *Mathematica* to find acceptable wavefunctions and energy levels for the hydrogen atom.

4.5   Hydrogen Atom

The Bohr Atom

In 1913, Niels Bohr formulated the quantum theory of the hydrogen atom. Bohr envisioned an electron circling the central nucleus like a planet about the sun. In a quantum formulation, only orbits with angular momentum $L = n\hbar$ would be allowed, thus only *certain energies* could be absorbed or emitted by the electron when changing orbits.

Let us now use the Bohr postulate $L = mvR = n\hbar$ to find the size of the Hydrogen atom and its energy levels. For a single electron held in orbit by the Coulomb force,

$$\frac{mv^2}{R} = \frac{Ze^2}{4\pi\epsilon_0 R^2} \qquad \text{and} \qquad v = \frac{Ze^2}{4\pi\epsilon_0\hbar n}$$

Since $L = mvR = n\hbar$,

$$R = \frac{n\hbar}{mv} = \frac{n^2\hbar^2 4\pi\epsilon_0}{Zme^2} = n^2\frac{a_0}{Z}$$

$$\text{where} \qquad a_0 = \frac{\hbar^2 4\pi\epsilon_0}{me^2} = .529177 \text{ Å} \quad \text{(the Bohr radius)}$$

Thus, for Hydrogen with $Z = 1$,

$$E_n = \frac{1}{2}mv^2 - \frac{Ze^2}{4\pi\epsilon_0 R} = -\frac{Z^2 e^2}{8\pi\epsilon_0 a_0 n^2} = \frac{-13.6057\,\text{eV}}{n^2}$$

The Schrodinger Equation for the Hydrogen Atom

In 1926, Erwin Schrodinger solved the following equation for the *radial wavefunction* $y[r]$

$$\frac{-\hbar^2}{2m}\left[\frac{2}{r}y'[r] + y''[r]\right] + V[r]\,y[r] = E\,y[r]$$

$$\text{where the potential is } V[r] = \frac{-Ze^2}{4\pi\epsilon_0 r} + \frac{\hbar^2}{2m}\frac{\ell(\ell+1)}{r^2}$$

to start, we shall choose $\ell = 0$ and find the wavefunctions, energy states, and the *average* radius, for $n = 1, 2, 3$ and compare the results to Bohr.

We use the following values for the fundamental constants

$$\hbar = 1.054572 \times 10^{-34}\text{ J}\cdot\text{s} \qquad m_e = 9.109389 \times 10^{-31}\text{ kg} \qquad \frac{\hbar^2}{2m} = 3.80998\text{ eV}\cdot\text{Å}^2$$

$$\epsilon_0 = 8.854188 \times 10^{-12}\text{C}^2/\text{N}\cdot\text{m}^2 \qquad e = 1.602177 \times 10^{-19}\text{ C} \qquad \frac{e^2}{4\pi\epsilon_0} = 14.3996\text{ eV}\cdot\text{A}$$

We write Schrodinger's equation as

$$\frac{2}{r}y'[r] + y''[r] = \frac{-2m}{\hbar^2}[E - V[r]]y[r]$$

$$\frac{2}{r}y'[r] + y''[r] = \frac{-2m}{\hbar^2}[E + \frac{Ze^2}{4\pi\epsilon_0 r}]y[r]$$

or, with energy E in eV and distance r in Angstroms

$$\frac{2}{r}y'[r] + y''[r] = -.262468[E + \frac{14.3996}{r}]y[r]$$

This is the equation we will solve with *Mathematica*.

First, however, we need to define the *boundary conditions.* In Schrodinger's theory for a zero angular momentum state $\ell = 0$, the wavefunction will have some value as $r \rightarrow 0$, and to be a bound state, the wavefunction will go to zero as $r \rightarrow \infty$. We will arbitrarily pick y[.0001] = 1 and y'[30] = 0.   In this way, we avoid the infinity at $r = 0$, and we also establish an *effective infinity* at large r. We will now choose some starting value of E (say, –20 eV) and increase the value of E until the wavefunction approaches 1 as $r \rightarrow 0$.

Now, with the wavefunction well-behaved, we may relax the inner boundary condition and choose y[0] = 1 and y'[30] = 0 at that energy *eigenvalue,* to get an ultra-precise value of the wavefunction.

Let's see how this plays out in practice: (refer to the graphics on next page) Say we choose E = –13.60 eV, the wavefunction heads to some extreme negative value as $r \rightarrow 0$
Say we choose E = –13.61 eV, the wavefunction goes to some extreme positive value as $r \rightarrow 0$
(Therefore the true energy eigenvalue will be somewhere midway between these two values.)

```
sol1 =
   DSolve[{2 * y'[r] + r * y''[r] == -.262468 * (-13.60 r + 14.3996) * y[r],
      y[0.0001] == 1, y'[30] == 0}, y[r], r]
Plot[Evaluate[y[r] /. sol1, {r, .000000001, .000001}]];
```

Out[81]= $\{\{y[r] \to e^{-1.88933\,r}$

$\qquad (2.19781\,\text{HypergeometricU}[-0.000205318, 2, 3.77866\,r] -$

$\qquad 8.23596 \times 10^{-42}\,\text{LaguerreL}[0.000205318, 1, 3.77866\,r])\}\}$

unacceptable behavior of the wavefunction $\psi \to -\infty$ **as** $r \to 0$ **E = -13.60 eV**

```
sol1 =
  DSolve[{2 * y'[r] + r * y''[r] == -.262468 * (-13.61 r + 14.3996) * y[r],
    y[0.0001] == 1, y'[30] == 0}, y[r], r]
Plot[Evaluate[y[r] /. sol1, {r, .000000001, .000001}]];
```

Out[116]=

$\{\{y[r] \to$

$\qquad e^{-1.89002\,r}\,(0.699228\,\text{HypergeometricU}[0.000162202, 2, 3.78005\,r] +$

$\qquad 3.17319 \times 10^{-42}\,\text{LaguerreL}[-0.000162202, 1, 3.78005\,r])\}\}$

unacceptable behavior of the wavefunction $\psi \to +\infty$ **as** $r \to 0$ **E = -13.61 eV**

The Shooting method as r → 0

The above two graphs tell us that the proper eigenvalue is between –13.60 and –13.61 eV. We may proceed by the *bisection method* to get an ever more accurate value. When we get close to an energy *eigenvalue*, then the graph of the wavefunction is well-behaved and y goes to a value near 1 as r→0.

On magnification, the slope is a straight-line.

```
sol1 =
 DSolve[{2 * y'[r] + r * y"[r] == -.262468 * (-13.605585 r + 14.3996) * y[r],
    y[0.0001] == 1, y'[30] == 0}, y[r], r]
Plot[Evaluate[y[r] /. sol1, {r, .001, 3}]];
Plot[Evaluate[y[r] /. sol1, {r, .000000001, .000001}]];
```

Out [18] = $\{\{y[r] \rightarrow e^{-1.88972\,r}\,(1.00019 -$
$$9.17501 \times 10^{-38}\,\text{LaguerreL}[8.18262 \times 10^{-9},\,1,\,3.77943\,r])\}\}$$

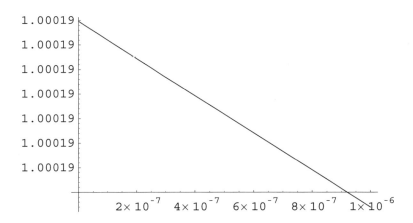

Acceptable behavior of the wavefunction as r → 0

After several trials, we find the energy eigenvalue E1 = - 13.605585 eV.

We will now run the *Mathematica* program one more time *this time with the proper initial conditions* y[0] = 1 and y'[30] = 0.

```
sol1 = DSolve[
    {2 * y'[r] + r * y"[r] == -.262468 * (-13.605585 r + 14.3996) * y[r],
    y[0] == 1, y'[30] == 0}, y[r], r] // Chop
Plot[Evaluate[y[r] /. sol1, {r, .001, 3}],
    Prolog → {Text["n=1, ℓ=0", {1.8, .56}]}];
Plot[Evaluate[r² * y[r]^2 /. sol1, {r, .001, 3}],
    Prolog → {Text["n=1, ℓ=0", {1.8, .026}]}];
```

Out [76] = {{y[r] → 1. e^{-1.88972 r}}}

We may now graph the wavefunction for $n = 1, \ell = 0$

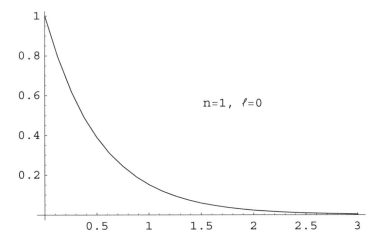

We may also evaluate $r^2 y^2[\, r\,]$ to find out where the probability is greatest of finding the electron

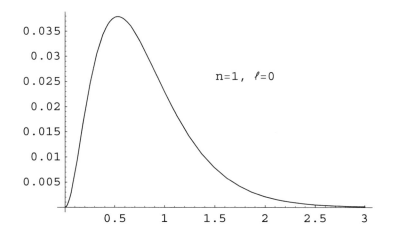

```
Table[{r, r² * y[r]^2 /. sol1}, {r, .51, .54, .001}] // TableForm
```

r (A)	r² y² [r]
0.51	0.0378471
0.511	0.0378523
0.512	0.0378573
0.513	0.037862
0.514	0.0378663
0.515	0.0378704
0.516	0.0378742
0.517	0.0378777
0.518	0.0378809
0.519	0.0378839
0.52	0.0378866
0.521	0.037889
0.522	0.0378911
0.523	0.0378929
0.524	0.0378944
0.525	0.0378957
0.526	0.0378967
0.527	0.0378975
0.528	0.0378979
0.529	0.0378981
0.53	0.037898
0.531	0.0378977
0.532	0.037897
0.533	0.0378961
0.534	0.037895
0.535	0.0378936
0.536	0.0378919
0.537	0.0378899
0.538	0.0378877
0.539	0.0378852
0.54	0.0378825

From the above table of values, we see that the maximum value of the *Probability function* $r^2 y^2$ occurs at r = 0.529 A or, more precisely, we maximize $r^2 (e^{-1.88972\,r})^2$

In[83]:= **Maximize[{r² Exp[-1.88972 r]^2, 0 < r < 10}, r]**

Out[83]= {0.037898, {r → 0.529179}}

which is the Bohr radius.

We will now find wavefunctions for n=2, ℓ=0 and n=3, ℓ=0 and n=2, ℓ=1. Using the Shooting Method as before,

```
soll = DSolve[
   {2/r * y'[r] + y''[r] == -.262468 * (-3.4013963256 + 14.3996/r) * y[r],
    y[0.0001] == 1, y'[30] == 0}, y[r], r] // Chop
Plot[Evaluate[y[r] /. soll, {r, .001, 7}], PlotRange → {-.15, 1.00},
   Prolog → {Text["n=2, ℓ=0", {2.3, .5}]}];
Plot[Evaluate[r^2 * y[r] ^2 /. soll, {r, .001, 7}],
   Prolog → {Text["n=2, ℓ=0", {5.4, .054}]}];
```

Out[179]= $\{\{y[r] \rightarrow e^{-0.944859\,r}\,(1.00019 - 0.945037\,r)\}\}$

We may now graph the wavefunction for n=2, ℓ=0

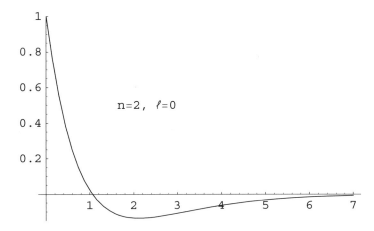

We evaluate $r^2 y^2[\,r\,]$ to find where the probability is greatest of finding the electron

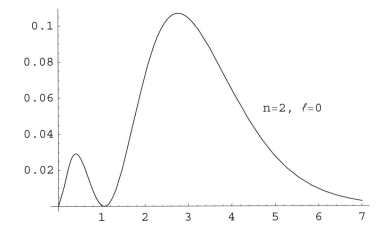

```
Clear[r, y];
soll = DSolve[
   {2 * y'[r] + r * y"[r] == -.262468 * (-1.51173167895 r + 14.3996) * y[r],
    y[0] == 1, y'[50] == 0}, y[r], r] // Chop
Plot[Evaluate[y[r] /. soll, {r, .001, 12}], PlotRange → {-.15, 1.00},
   Prolog → {Text["n=3, ℓ=0", {3.3, .5}]}];
Plot[Evaluate[r² * y[r]^2 /. soll, {r, .001, 12}],
   Prolog → {Text["n=3, ℓ=0", {3.2, .11}]}];
```

Out[201]= $\{\{y[r] \to e^{-0.629906\,r}\,(1. - 1.25981\,r + 0.264521\,r^2)\}\}$

The wavefunction for n=3, ℓ=0

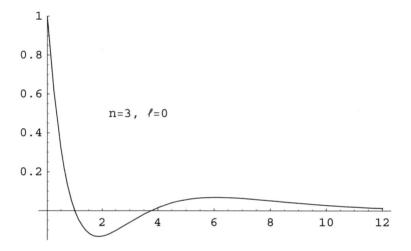

The probability function $r^2y^2[\,r\,]$ for n=3, ℓ=0

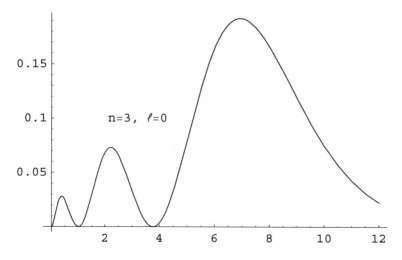

For n = 2, ℓ = 1, we need to revisit the original Schrodinger equation for the hydrogen atom.

$$\frac{2}{r}y'[r] + y''[r] = -.262468\,[\,E + \frac{14.3996}{r}\,]\,y[\,r\,] + \frac{\ell\,(\ell+1)}{r^2}y[\,r\,]$$

We will solve this equation using *Mathematica* with the proviso that for any $\ell > 0$ angular momentum state that y[r]$\to$ 0 as r $\to$ 0. It is amazing, but true, that the energy eigenvalue for the n =2, ℓ=1 state is the same as the n =2, ℓ=0 state. We show this in the expanded graphic where the wavefunction y approaches zero as r $\to$ 0.

```
DSolve[{ 2/r *y'[r] +y"[r] == -.262468*(-3.4013963256 + 14.3996/r )*y[r]

   + 1 (1+1)/r^2 *y[r], y[0.0001] == 0.00001, y'[50] == 0}, y[r], r] // Chop
Plot[Evaluate[y[r] /. %, {r, .000000001, .000001}]];
```

Out[7]= {{y[r] → 0.100009 e^{-0.944859 r} r^{1.}}}

We may now evaluate the wavefunction and the probability function for n =2, ℓ =1

```
sol1 = DSolve[{ 2/r *y'[r] +y"[r] ==

   -.262468*(-3.4013963256 + 14.3996/r )*y[r] + 1 (1+1)/r^2 *y[r],
   y[0.000001] == 0.0000001, y'[50] == 0}, y[r], r] // Chop
Plot[Evaluate[y[r] /. sol1, {r, .001, 7}],
   Prolog → {Text["n=2, ℓ=1", {4.8, .022}]}];
Plot[Evaluate[r^2 *y[r]^2 /. sol1, {r, .001, 7}],
   Prolog → {Text["n=2, ℓ=1", {5, .0017}]}];
```

Out [33] = {{y[r] → 0.1 e$^{-0.944859 r}$ r$^{1.}$}}

The wavefunction for n=2, ℓ=1

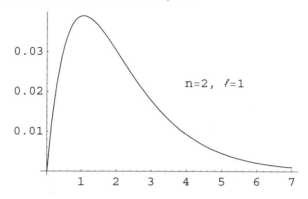

n=2, ℓ=1

The probability function $r^2 y^2$[r] for n=2, ℓ=1

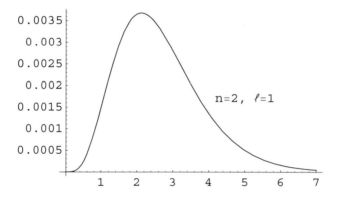

n=2, ℓ=1

It is now time to normalize all the wavefunctions

For n=1, ℓ=0,   y[r] = A e$^{-1.88972 r}$
For n=2, ℓ=0,   y[r] = A e$^{-0.944859 r}$ (1. $-$ 0.944859 r)
For n=2, ℓ=1   y[r] = A e$^{-0.944859 r}$ r
For n=3, ℓ=0   y[r] = A e$^{-0.629906 r}$ (1. $-$ 1.25981 r + 0.264521 r^2) }}

(*n=1,l=0*) Solve$\left[A^2 \int_0^\infty r^2 * \text{Exp}[-1.88972 * 2 * r] \, dr == 1, A \right]$

(*n=2,l=0*) Solve$\left[A^2 \int_0^\infty r^2 * (1 - .944859 \, r)^2 \text{Exp}[-.944859 * 2 * r] \, dr == 1, A \right]$

(*n=2,l=1*) Solve$\left[A^2 \int_0^\infty r^2 * r^2 \text{Exp}[-.944859 * 2 * r] \, dr == 1, A \right]$

(*n=3,l=0*)

Solve$\left[A^2 \int_0^\infty r^2 * (1 - 1.25981 \, r + .264521 \, r^2)^2 \text{Exp}[-.629906 * 2 * r] \, dr == 1, A \right]$

Out [1] = {{A → 5.19549}}   {A → 1.83688}   {A → 1.00204}   {A → 0.9998689}}

To summarize, the first four radial wavefunctions for the hydrogen atom, and their energy levels in terms of the Bohr radius $a_0 = 0.529179$ Å are

$n=1, \ell=0$ $y_{10}[r] = 5.19549 \; e^{-r/a_0}$ $\qquad\qquad\qquad$ $E_1 = -13.6056 \, \text{eV}$

$n=2, \ell=0$ $y_{20}[r] = 1.83688 \; (1-\frac{r}{2\,a_0}) \; e^{-r/2\,a_0}$ $\qquad$ $E_2 = -3.4014 \, \text{eV}$

$n=2, \ell=1$ $y_{21}[r] = 1.00204 \; r \; e^{-r/2\,a_0}$ $\qquad\qquad$ $E_2 = -3.4014 \, \text{eV}$

$n=3, \ell=0$ $y_{30}[r] = 0.99987 \; (1-\frac{2\,r}{3\,a_0}+\frac{2\,r^2}{27\,a_0{}^2}) \; e^{-r/3\,a_0}$ $E_3 = -1.5117 \, \text{eV}$

It is worth noting that the Niels Bohr theory, when corrected for the finite mass of the nucleus, and the relativistic mass increase of the electron ($v = .0073$ c in the ground state), agrees with spectroscopic data to 3 parts in 100,000. The Bohr theory was ground-breaking for its time because it showed that atoms could only be described in quantum terms.

However, the Bohr theory only works for hydrogen and singly-ionized helium atoms. Schrodinger's theory predicts exactly the same results as Bohr for hydrogen, and when the Dirac relativistic theory of the electron is applied, corresponds exactly with the hydrogen spectrum.

The value of the Schrodinger theory is that it may be applied to all quantum-mechanical systems.

A question that we should answer here is how well the above "Shooting Method" compares with the exact quantum mechanical solution for the Hydrogen atom.

As given in Leonard Schiff's *Quantum Mechanics* (1968) or in Harald Enge's *Introduction to Atomic Physics* (1972) the exact quantum mechanical wavefunctions for hydrogen, which we evaluate with $Z=1$ and $a_0 = 0.529179$ Å are

$n=1, \; \ell=0$ $y_{10}[r] = \left(\frac{Z}{a_0}\right)^{3/2} 2 \, e^{-Zr/a_0}$ $\qquad\qquad = 5.19549 \; e^{-r/a_0}$

$n=2, \; \ell=0$ $y_{20}[r] = \left(\frac{Z}{2\,a_0}\right)^{3/2} 2 \, (1-\frac{Zr}{2\,a_0}) \; e^{-Zr/2\,a_0} = 1.83688 \; (1-\frac{r}{2\,a_0}) \; e^{-r/2\,a_0}$

$n=2, \; \ell=1$ $y_{21}[r] = \left(\frac{Z}{2\,a_0}\right)^{3/2} \frac{Zr}{\sqrt{3}\,a_0} \; e^{-Zr/2\,a_0} \qquad = 1.00205 \; r \; e^{-r/2\,a_0}$

$n=3, \; \ell=0$ $y_{30}[r] = \left(\frac{Z}{3\,a_0}\right)^{3/2} 2 \, (1-\frac{2\,Zr}{3\,a_0}+\frac{2\,Z^2\,r^2}{27\,a_0^2}) \; e^{-Zr/3\,a_0}$
$$= 0.99987 \; (1-\frac{2\,r}{3\,a_0}+\frac{2\,r^2}{27\,a_0^2}) e^{-r/3\,a_0}$$

The exact wavefunctions and those derived via the Shooting Method correspond to 6-decimal places. Any very small inaccuracy is due to our choice of an "effective infinity" for the wavefunction to go to zero at large r. Thus the choice of the Shooting Method will allow evaluation of the energy eigenvalues and the wavefunctions for potentials where Schrodinger's equation cannot be solved analytically.

4.6   Deuterium

Deuterium is the simplest nuclear system. Here we have one proton bound to one neutron to form the nucleus of heavy hydrogen. Let us assume the attractive potential between the two particles is 40 MeV acting to some distance a. Let us find the wavefunction for Deuterium, given that the energy needed to break apart (photodissociate) this nucleus is only 2.225 MeV.

```
Plot[{.566738 * Sin[.953 r] * If[r < 1.898, 1, 0] +
   .85384 * Exp[-.232 r] * UnitStep[r - 1.9],
  -.4 * UnitStep[r] + .4 * UnitStep[r - 1.9]}, {r, 0, 4},
 PlotRange → {-0.42, 0.58}, PlotStyle → {GrayLevel[0], GrayLevel[0.45]}]
```

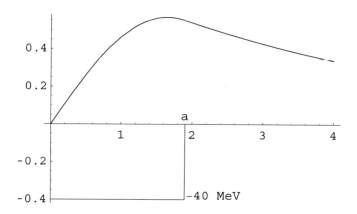

The wavefunction for Deuterium is barely anchored in the 40-MeV well.

Schrodinger's wave equation in center-of-mass coordinates is

$$\frac{-\hbar^2}{2m} \psi'' + V(r)\psi = E\psi$$

with solution $\psi = A \sin kr$ for $0 < r < a$

$\psi = B\, e^{-\alpha r}$ for $r \ge a$

Using the reduced mass for *m* and the binding energy –2.225 MeV for E,

$$k = \sqrt{1.673 * 10^{-27} * (40 - 2.225) * 1.6 * 10^{-13}} \Big/ (6.626 * 10^{-34} / (2\pi)) \Big/ 10^{15}$$

$$\alpha = \sqrt{1.673 * 10^{-27} * 2.225 * 1.6 * 10^{-13}} \Big/ (6.626 * 10^{-34} / (2\pi)) \Big/ 10^{15}$$

$$k = 0.953539 \quad \alpha = 0.23142 \quad \text{in units of Fm}^{-1}$$

We may now find the range of the 40 MeV attractive potential that produces –2.225 MeV binding energy. Match the wavefunction components and their derivatives at $r = a$.

$$A \operatorname{Sin} ka = B \, e^{-\alpha a} \qquad \text{and}$$

$$kA \operatorname{Cos} ka = -\alpha \, B \, e^{-\alpha a} \qquad \text{at } r = a$$

Dividing the first equation by the second gives us an equation in **a**

```
In[7]:=  FindRoot[Tan[k*a] == - k/α, {a, 2, 1, 3}]
```

```
Out[7]= {a → 1.89873}
```

We may now find **A** and **B** from the normalization condition and the continuity of the wavefunction at **a**

```
In[6]:=  a = 1.898; k = .953; α = .231;
```

$$\texttt{FindRoot}\Big[\Big\{1 == A^2 \int_0^a (\operatorname{Sin}[k*r])^2 \, dr + B^2 \int_a^\infty \operatorname{Exp}[-2 \, \alpha*r] \, dr,$$

$$A*\operatorname{Sin}[k*a] == B*\operatorname{Exp}[-\alpha*a]\Big\}, \{\{A, .5\}, \{B, .5\}\}\Big]$$

```
Out[8]= {A → 0.566738, B → 0.853839}
```

Let's find the wavefunction probability ψ^2 of finding the particle within the radius **a** of the attractive potential

```
In[9]:=  A = 0.566738; B = 0.853839;
```

$$\texttt{Pr1} = A^2 \int_0^a (\operatorname{Sin}[k*r])^2 \, dr \qquad \texttt{Pr2} = B^2 \int_a^\infty \operatorname{Exp}[-2 \, \alpha*r] \, dr$$

```
Out[10]=   Pr1 = .34342         Pr2 =   0.65658
```

This says that the Deuteron only spends 34% of the time entirely within the range of the attractive nuclear potential. The Deuteron is more of a wave than a particle.

Let's now find the *effective radius* of the Deuteron.

```
In[12]:=
```
$$\texttt{rad} = A^2 \int_0^a r*(\operatorname{Sin}[k*r])^2 \, dr + B^2 \int_a^\infty r*\operatorname{Exp}[-2 \, \alpha*r] \, dr$$

```
Print[rad, " Fermis"]
```

```
3.1134 Fermis
```

Which considerably exceeds the range of the attractive nuclear potential. The wavefunction for the Deuteron is barely anchored in the 40-MeV well.

4.7 Challenge Problems

1. Ramsauer Effect

An interesting quantum-mechanical effect is observed if low-velocity electrons are sent into a gas of Xenon, Argon, or Krypton. Depending on the energy of the electrons, they will not be scattered at certain energies, rather they will transmit through the gas. As a simple model, assume that surrounding each gas nucleus there is an attractive square-well potential of -8.4 eV depth and width 2 Angstroms. What is the transmission T for 1 eV electrons?

2. Intermediate Quantum Step

We can get 100 % transmission across a potential step if we put in an intermediate quantum step of the appropriate height V and width L. For a 7-eV electron beam impinging on a 5-eV potential step, how high and how wide does the intermediate step need to be? Express the answer in eV and Angstroms.

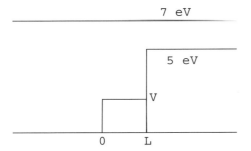

Intermediate Quantum Step

3. Energy Eigenvalues

An electron is trapped in a quantum well. If this is a square well of depth 5 eV and width 2 Angstroms, what is the ground state energy of the electron above the bottom of the well? The ground-state wavefunction for the electron will be a Cosine wave, because this allows a maximum probability at the center of the well. Notice that it is necessary to match the Cosine wave in the well with an exponentially decreasing wavefunction outside the well.

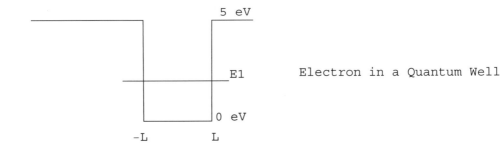

Electron in a Quantum Well

4. Cooling Concrete in Dams

The dams on the Columbia River are Grand Coulee, McNary, John Day, and Bonneville. Each one may appear to be a quiescent object holding back millions of tons of water. It is inside the dam that all the activity is: the flow of water through Grand Coulee powers generators that provide over 6000 MW of electrical power, and even inside the walls there is heat retained from the formation of the dam.

To show how long it takes the concrete to equilibrate to the outside temperature, let us consider a 20-foot-wide slab of concrete. When freshly poured, as chemical reactions proceed toward solidification, the concrete may have a temperature of 130 °F. Let us assume the temperature of the slab is constant throughout at t = 0 and we further assume that the outside temperature is constant at 60 °F on both walls of the slab. In this way, we may model the cooling of the concrete as a one-dimensional heat-flow problem

$$\frac{\partial T}{\partial t} = \alpha \frac{\partial^2 T}{\partial x^2} \qquad \text{where} \qquad \alpha = \frac{k}{\rho C}$$

Given the following thermal properties of concrete,

thermal conductivity $k = 1.5$ BTU/hr/ft²·°F, density $\rho = 150$ lb/ ft³,

and specific heat $C = 0.25$ BTU/ °F·lb,

What is the temperature in the center of the slab after 40 days?

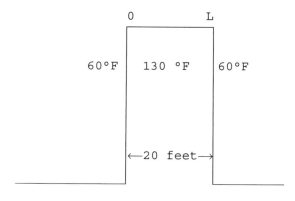

concrete slab at t = 0

5. Heat Transfer in Steel

A thick steel slab at 550 °F has its surface suddenly cooled to 100 °F. How long before the temperature at 1-inch depth reaches 200 °F ?

The Partial Differential Equation for heat transfer in steel is given by

$$\frac{\partial T}{\partial t} = \alpha \frac{\partial^2 T}{\partial x^2} \qquad \text{where} \qquad \alpha = 0.45 \text{ ft}^2 / \text{hr}$$

Chapter 5 Applications

5.1   Laser Pulse Dynamics

In their 1988 book, *Lasers*, Peter Milonni and Joseph Eberly model the generation of a laser pulse as a collapse of a population inversion of electrons in a higher energy state. If the laser is "pumped-up" so that twice as many electrons are in an excited state, relative to the lower energy state, then as a few of the electrons de-excite, more and more photons are generated *in phase and with the same energy.* So, no matter how few photons were originally in the laser, the output signal is increased exponentially until the upper energy state is unloaded. This process will occur over several nanoseconds, and may be modeled by the following equations: (We set x equal to the intensity of the beam, and y as the population inversion)

$$x' = (y-1)*x \quad \text{and} \quad y' = -x*y \quad \text{where} \quad x_o = .0001 \text{ and } y_o = 2$$

Notice that no laser amplification is achieved if $x_o = 0$. If the starting value of x is zero, then the first equation says that x' can never increase.

These equations are easy to program in *Mathematica.*

```
sol = NDSolve[{x'[t] == (y[t] - 1) *x[t],
    y'[t] == -x[t] *y[t], x[0] == .0001, y[0] == 2}, {x, y}, {t, 0, 20}]
InterpFunc1 = x /. sol[[1]]; InterpFunc2 = y /. sol[[1]];
Plot[Evaluate[{10 *x[t], y[t]} /. sol, {t, 0, 20}],
    Epilog → {Text["Laser" , {13, 2.5}], Text["Intensity", {13.5, 2.25}],
        Text["y, 10x", {1.8, 3}], Text["Inversion", {17.5, .6}]}];
```

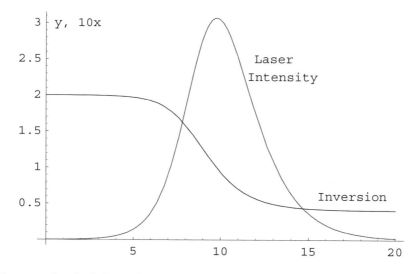

Laser pulse height and population inversion vs time in nanoseconds

An interesting question is "How much power is contained in this laser pulse?" Following Milloni and Eberly, (*Lasers*, Chapter 12) we use these values for x and y

$$x = \frac{I}{ch\nu\, N_t} \quad \text{and} \quad y = \frac{N_2}{N_1}$$ where N_t for ruby is 3.7×10^{17} atoms/cm³ $\lambda = 6943$ A.

Then, the peak intensity is given by $I = (c\, h\, \nu)\, N_t\, (0.3) \approx 10^9$ Watts/cm² in the laser beam. This very large number is due to the fact that we have overloaded the upper population level with approximately 10^{18} excited electrons. This can be accomplished in most lasers by delaying the onset of the laser "download" by making the cavity of the laser *non-resonant.*

This allows the excited-electron population to build to an extremely high level, such as $y = 2$. The laser cavity is then turned *resonant* at $t = 0$, and the laser pulse builds in intensity and then escapes. This process of allowing the laser to build in power is called Q-switching.

```
tbl = Table[{t, InterpFunc1[t], InterpFunc2[t]}, {t, 5, 15, .5}] //
    TableForm
```

t (ns)	Intensity	Inversion
5	0.0144311	1.97113
5.5	0.0233538	1.95293
6.	0.0373577	1.924
6.5	0.0586797	1.87903
7.	0.0896394	1.81143
7.5	0.131455	1.71479
8.	0.182211	1.58588
8.5	0.235043	1.42861
9.	0.278974	1.25556
9.5	0.303583	1.08436
10.	0.304422	0.930539
10.5	0.284446	0.802567
11.	0.25102	0.701738
11.5	0.21194	0.624984
12.	0.173081	0.567687
12.5	0.137892	0.525328
13.	0.107872	0.494135
13.5	0.0832608	0.471183
14.	0.0636328	0.454281
14.5	0.048279	0.441819
15.	0.0364335	0.432617

5.2   The B–Z Chemical Reaction

The Belousov-Zhabotinsky chemical reaction

A series of chemical reactions using bromates to oxidize Cerium or Manganese will under proper conditions lead to an oscillating chemical reaction. The color of a solution will change from green to blue to violet and red then repeat the sequence of colors as many as 20 times. This curious reaction was discovered by Boris Belousov in 1951.

The reaction was shown to occur in several other chemical systems by Anatol Zhabotinsky in 1958. The sequence of colors is due to a complicated set of color-absorbing ions whose concentrations change in a periodic manner as the reaction proceeds.

The differential equations of the reaction are (using x, y, and z as the different interacting ions) as follows:

$$x' = 15\,(.0005y - xy + x - x^2)$$
$$y' = 2500\,(-.0005\,y - xy + z)$$
$$z' = \ x - z$$

with initial conditions $x(0) = 1,\ y(0) = 0,\ z(0) = 0$

Because of the vastly different concentrations of x, y, and z, and their remarkably fast rates of change, this is a STIFF DIFFERENTIAL EQUATION. It is to *Mathematica*'s credit that the adaptive algorithm used in **NDSOLVE** is able to test for stiffness, and then solve the reaction equations with mathematical accuracy.

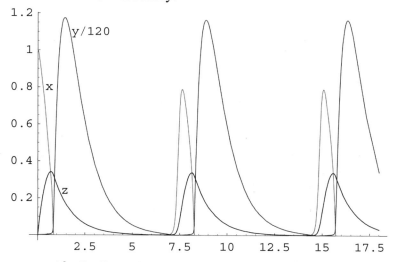

The B – Z reactants : x = green, y = red, z = blue

```
In[1]:=  sol =
           NDSolve[{x'[t] == 15 (.0005 y[t] - (x[t] * y[t]) + x[t] - x[t]^2),
              y'[t] == 2500 (-.0005 y[t] - x[t] * y[t] + z[t]),
              z'[t] == x[t] - z[t], x[0] == 1, y[0] == 0,
              z[0] == 0}, {x, y, z}, {t, 0, 18}]

In[2]:=  InterpFunc1 = x /. sol[[1]]; InterpFunc2 = y /. sol[[1]];
           InterpFunc3 = z /. sol[[1]];
           Chop[Table[{t, InterpFunc1[t], InterpFunc2[t], InterpFunc3[t]},
              {t, 0, 8, .4}] // TableForm]
           Plot[Evaluate[{x[t], y[t] / 150, z[t]} /. sol],
              {t, 0, 18}, Prolog → {Text["x", {.6, .8}],
                 Text["y/120", {2.8, 1.1}], Text["z", {1.44, .24}]}]
```

t (s)	x	y	z
0	1.	0	0
0.4	0.685862	0.409051	0.281224
0.8	0.076063	3.69684	0.328543
1.2	0.000503115	161.497	0.221106
1.6	0.000502946	170.611	0.148378
2.	0.000503665	137.303	0.0996265
2.4	0.000505021	100.475	0.0669479
2.8	0.000507189	70.4543	0.0450434
3.2	0.000510536	48.3675	0.0303612
3.6	0.000515663	32.8393	0.0205209
4.	0.000523536	22.1636	0.0139269
4.4	0.00053575	14.9075	0.00951006
4.8	0.000555046	10.0043	0.00655455
5.2	0.000586499	6.69879	0.00458173
5.6	0.000640594	4.4684	0.00327323
6.	0.000743342	2.95276	0.00242138
6.4	0.000984731	1.89301	0.00190396
6.8	0.00202584	1.03856	0.00173055
7.2	0.0500508	0.106964	0.00519576
7.6	0.786846	0.195808	0.154469
8.	0.510139	0.636694	0.325821

Notice that after the initial reactions where x, y, and z all peak, the reactants almost entirely die out before undergoing a resurgence just after $t = 7.2$ s. Mathematically, this is a consequence of the vastly different coefficients in the differential equations. Chemically, this is due to very rapid rates in transforming the reaction products into other intermediates which themselves transform, and reappear in the original reactions. So the original reactants that are destroyed in the earlier reactions reappear in later reactions, and the process oscillates back-and-forth.

```
In[38]:= ParametricPlot3D[Evaluate[{x[t], y[t]/150, z[t]} /. sol],
         {t, 0, 10}, PlotRange → {{-.5, 1}, {-.5, 1.5}, {-.5, 1}},
         PlotPoints → 500, AxesLabel → {"x", "y", "z"},
         Prolog → Text["(1,0,0)", {.75, .225}]]
```

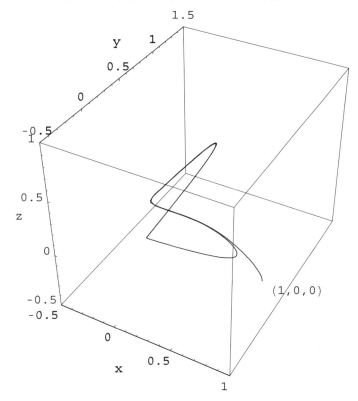

The time track of the concentrations x, y, z in the B-Z reaction.
After t = 2 s, the figure-8 curve traces itself over and over.

Starting at $x = 1$, $y = 0$, $z = 0$ at $t = 0$, the curve decreases in x and increases along z until it makes a sharp turn to a very-high value of y after $t = 0.8$ s. Then the curve tracks down in z and decreases in y as it approaches 0,0,0 near $t = 6.8$ s. Then the curve begins to increase again in x, and traces the distorted figure-8 track over and over as the cycle repeats. Notice that the y-axis is compressed by a factor of 150.

In[3]:=

```
sol = NDSolve[{x'[t] == 15 (.0005 y[t] - (x[t] *y[t]) + x[t] - x[t]^2),
    y'[t] == 2500 (-.0005 y[t] - x[t] *y[t] + z[t]), z'[t] == x[t] - z[t],
    x[0] == 1, y[0] == 0, z[0] == 0}, {x, y, z}, {t, 0, 12}]

Module[{gridSpec},
 gridSpec = Table[{k, GrayLevel@0.65}, {k, -0.5, 1, 0.5}];
 ParametricPlot3D[Evaluate[{x[t], y[t] / 150, z[t]} /. sol],
  {t, 0, 10}, PlotRange -> {{-.5, 1}, {-.5, 1.5}, {-.5, 1}},
  PlotPoints -> 500, AxesLabel -> {"x", "y", "z"},
  FaceGrids -> {{{0, 0, -1}, {gridSpec, gridSpec}},
    {{0, 1, 0}, {gridSpec, gridSpec}}},
  ViewPoint -> {1.446, -2.952, 0.803}];]
```

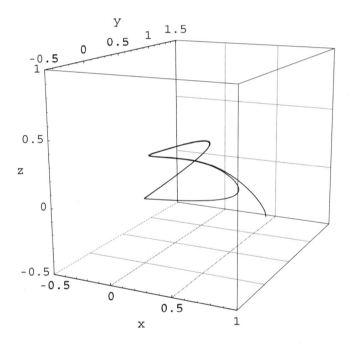

With grid-lines and a different perspective.
After $t = 2$ s, the figure-8 curve traces itself over and over.

5.3   Foxes and Rabbits

Vito Volterra in the 1930s wrote down the equations for the interaction of two species, one a predator (foxes) and their prey (rabbits). Since the foxes depend upon the rabbits as food, the two populations change in time – if there is a sufficient number of rabbits, the foxes will increase in time, however if there are too few rabbits, then the fox population will decrease in time.

Let us take a somewhat simplified situation, where we let loose some rabbits into a very large grassland. We let their numbers increase to 1000, and then we introduce 8 foxes.

We will plot the number of rabbits x and foxes y versus time, and then show that the two populations are <u>periodic</u> in time, and find the time for the cycle to repeat.

With time measured in days, the Predator-Prey equations are

for rabbits $\dot{x} = \alpha\, x - \beta\, xy$ $x_0 = 1000$

for foxes $\dot{y} = -\epsilon\, y + f\, xy$ $y_0 = 8$

where $\beta = 0.001$ (each fox averages one rabbit per day when x = 1000)

  $\alpha = 0.01$ (the rabbit population initially increases at 10 rabbits per day)

  $\epsilon = 0.015$ (the fox mortality rate if rabbits arent caught)

  $f = \beta/40$ (the conversion efficiency of rabbits into new foxes)

Mathematica easily solves the Predator-Prey equations

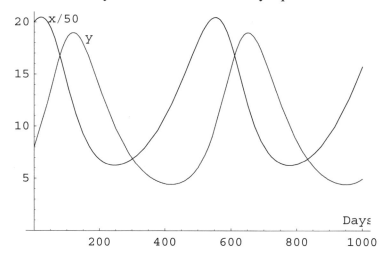

Number of rabbits x / 50 and number of foxes y versus time in days

```
In[17]:=
Clear[α, β, ε, f, x, y, t]; α = .01; β = .001; ε = .015; f = β / 40;
sol = NDSolve[{x'[t] == α * x[t] - β * x[t] * y[t],
    y'[t] == -ε * y[t] + f * x[t] * y[t],
    x[0] == 1000, y[0] == 8}, {x, y}, {t, 0, 1000}]
InterpFunc1 = x /. sol[[1]]; InterpFunc2 = y /. sol[[1]];
InterpFunc3 = x' /. sol[[1]]; InterpFunc4 = y' /. sol[[1]];
Plot[Evaluate[{x[t] / 50, y[t]} /. sol, {t, 0, 1000}], PlotRange → {0, 22},
    Prolog → {Text["x/50", {90, 20.3}], Text["y", {168, 18.5}]},
    PlotStyle → {{RGBColor[0, 0, 1]}, {RGBColor[1, 0, 0]}}];
Table[{t, InterpFunc1[t], InterpFunc2[t]}, {t, 0, 540, 30}] //
    TableForm
```

If we **Table** the results, we will find the <u>maximum</u> number of foxes and the <u>minimum</u> number of rabbits. We see that the number of foxes increases to about 19 and the rabbit population falls to about 310. (The rabbits dont do too well on fox-rabbit encounters). However, as the number of rabbits decreases, the fox population must also fall (food supply almost gone) so then as the number of predators decreases, the rabbits begin to increase, and the cycle begins again. The **Table** or the graph shows that the ecological game cycle repeats about every 530 days.

t (days)	x (rabbits)	y (foxes)
0	1000.	8.
30	1018.76	10.9358
60	938.01	14.6178
90	776.128	17.7836
120	600.05	18.9738
150	463.246	17.9678
180	376.912	15.6537
210	331.091	12.9886
240	314.341	10.5336
270	319.266	8.50771
300	342.19	6.94452
330	381.889	5.80337
360	438.566	5.02802
390	512.977	4.57554
420	605.404	4.43271
450	713.992	4.63164
480	831.82	5.27189
510	942.229	6.54639
540	1014.11	8.72517

Another way of looking at this ecological game cycle is to plot the number of rabbits x versus the number of foxes y. In a **Parametric Plot**, we can see how the two populations move along a closed path.

```
In[24]:= ParametricPlot[{x[t], y[t]} /. sol,
           {t, 0, 550}, Prolog → Text["Δ", {1000, 8}],
           PlotRange → {{300, 1040}, {4, 19}},
           AxesLabel → {"x rabbits", "y foxes"}];
```

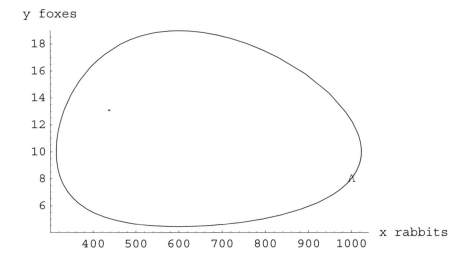

The number of rabbits and the number of foxes as a function of time in days. The cycle starts at Δ (1000, 8) and moves counter-clockwise.

5.4 PDE – Nerve Conduction

Nerve Conduction

In 1952, after more than 20 years of work, Alan Hodgkin and Andrew Huxley published a theory of nerve conduction. (Journal of Physiology, 117, 500-544).

The first thing that Hodgkin and Huxley strove to understand was how a nerve fiber with an electrical resistance of 100 million ohms could conduct electrical signals. What they found, by using radioactive tracer elements was that the inside of a nerve cable (or nerve *axon*) has an abundance of potassium ions and outside the nerve axon is a super-abundance of sodium ions.

When the nerve receives a stimulus at the end, the permeability of the membrane that surrounds the nerve changes, allowing sodium (Na$^+$) in and potassium (K$^+$) to surge out. Hodgkin-Huxley found that the applied voltage changed the rate at which the conductivity of the membrane changed.

Thus there is a feedback mechanism whereby the initiating voltage causes a current flow through the nerve membrane which changes the voltage, which allows current flow ever further down the line. This is the *action potential*, a 100-mV voltage pulse that moves down the nerve axon at a rate determined by the capacitance of the nerve membrane, and the rate at which the conductivity changes.

By a series of very careful measurements on nerve fiber in squid, Hodgkin and Huxley were able to determine how the conductivity of the nerve membrane changed as a function of voltage.

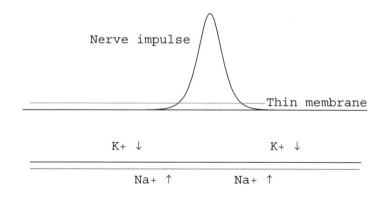

The nerve axon with potassium flow outward
and sodium flow inward produces the nerve impulse

The Voltage-Clamp experiments

To determine how the conductivity of the nerve changes with voltage, Hodgkin and Huxley placed a fine wire down the center of the axon, and then applied a current impulse or voltage step. In this way, the potassium and sodium conductivity could be determined as a function of voltage and the nerve impulse would occur at the same time all along the axon. The expressions they found for the conductivity are as follows:

The potassium conductivity is $g_K = .036 * n^4$

The sodium conductivity is $g_{NA} = .120 * m^3 h$

where n, m, and h are found to change with time as

$$n' = \alpha (1 - n) - \beta n \qquad m' = \gamma (1 - m) - \delta m \qquad h' = \epsilon (1 - h) - \eta h$$

and α, β, γ, δ, ϵ, η are functions of voltage and at 6.3 °C are given by

$$\alpha = \frac{.01 * (10 - V)}{Exp[(10 - V)/10] - 1} \qquad \beta = 0.125 * Exp[-\frac{V}{80}]$$

$$\gamma = \frac{10 * (25 - V)}{Exp[(25 - V)/10] - 1} \qquad \delta = 4.0 * Exp[-\frac{V}{18}]$$

$$\epsilon = .07 * Exp[-\frac{V}{20}] \qquad \eta = \frac{1}{Exp[(30-V)/10] + 1}$$

These correspond to the opening and closing of channels in the protein structure of the nerve membrane as the voltage changes.

We are now ready to put together the *conductivity equations* that show how the current flows through the nerve membrane. The current due to the potassium and sodium ion flow is

$$i_{ion} = g_K (V + 12) + g_{NA} (V - 115) + g_L (V - 10.6) \quad \text{or}$$

$$i_{ion} = .036 * n^4 (V + 12) + .120 * m^3 h (V - 115) + .0003 (V - 10.6)$$

where g_L is a leakage term for all other ions, like Calcium, that keep the ion current at zero when the nerve is at rest. The current is measured in mA, the voltage is in mV. We now add in the capacitance of the membrane and the initial current to obtain the H–H *voltage-clamp equations*

$$i_{init} = C_m \frac{dV}{dt} + .036 * n^4 (V + 12) + .120 * m^3 h (V - 115) + .0003 (V - 10.6)$$

We will apply a 0.1 mA initial current pulse to the nerve and we will see how the voltage changes in time as sodium surges in and potassium surges out.

Again, V is measured from its rest potential of zero, and is in millivolts. $C_m = 0.001$ mF is the membrane capacitance, and time is measured in milliseconds. Using *Mathematica*,

```
Clear[V, z, m, h, n];
vsol1 = NDSolve[{.001 * V'[t] + .036 * n[t]^4 * (V[t] + 12) +
     .120 * m[t]^3 * h[t] * (V[t] - 115) + .0003 * (V[t] - 10.6) ==
  0.1 * Which[t < .3, 0, .3 ≤ t ≤ .4, 1, t > .4, 0], n'[t] ==
     .01 * (10 - V[t])
  ─────────────────── * (1 - n[t]) - .125 * Exp[- V[t] ] * n[t],
  Exp[(10 - V[t]) / 10] - 1                          80

        .1 * (25 - V[t])
m'[t] == ─────────────────── * (1 - m[t]) - 4.0 * Exp[- V[t] ] * m[t],
        Exp[(25 - V[t]) / 10] - 1                        18

h'[t] == .07 * Exp[- V[t] ] * (1 - h[t]) -       1        * h[t],
                    20                     Exp[(30 - V[t]) / 10] + 1
n[0] == .3177, m[0] == .0529, h[0] == .5961, V[0] == 0},
  {V, n, m, h}, {t, 0, 4.2}]
Plot[Evaluate[V[t] /. vsol1], {t, 0, 3.9},
 PlotRange → {{0, 4.2}, {0, 105}}, Prolog →
 {Text["100-mV", {3.1, 82}], Text["Action Potential", {3.4, 72}],
   Text["mV", {0.18, 101}], Text["t(ms)", {4.0, 6.7}]}]
InterpFunc1 = V /. vsol1[[1]]; InterpFunc2 = n /. vsol1[[1]];
InterpFunc3 = m /. vsol1[[1]]; InterpFunc4 = h /. vsol1[[1]];
V1[t_] = InterpFunc1[t]; n1[t_] = InterpFunc2[t];
m2[t_] = InterpFunc3[t]; h3[t_] = InterpFunc4[t];
 Plot[{36 * n1[t]^4, 120 * m2[t]^3 * h3[t]}, {t, 0, 4.0}]
 Plot[0.1 (UnitStep[t - .3] - UnitStep[t - 0.4]), {t, 0, 2.0}]
```

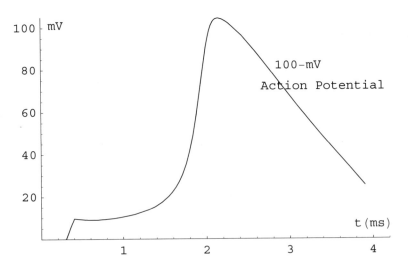

In the voltage-clamp experiments, the action potential arises at all points along the axon.

It is interesting to compare the 100-mV *action potential* to the sodium and potassium conductance and also to the rather small current of 0.1 mA for 0.1 ms that triggered the 100-mV voltage response all along the voltage-clamped nerve axon. It is the rapid increase in sodium conductance that starts the *action potential* and the steady increase of the potassium conductance that brings the action potential back to zero.

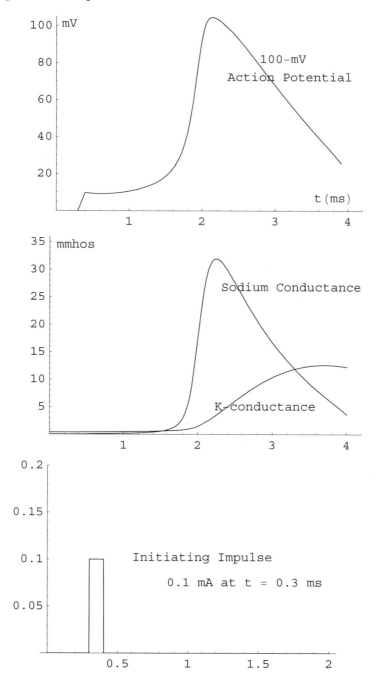

The Full Hodgkin–Huxley Equations of Nerve Conduction

The next question is whether the H–H model of nerve conduction can account for the actual speed of propagation of the *action potential* down the nerve axon. For this, Hodgkin and Huxley used the Partial Differential Equation of cable conduction with additional terms for sodium and potassium conductance. The PDE is

$$\frac{a}{2\rho} \frac{\partial^2 V}{\partial x^2} = C_m \frac{\partial V}{\partial t} + .036\, n^4\, (V + 12) + .12\, m^3\, h\, (V - 115) + .0003\, (V - 10.6)$$

where a = 0.0238 cm is the radius of the axon, and ρ = 35.4 Ω-cm is the resistivity of the membrane. Hodgkin-Huxley actually solved this equation using only a desk calculator! But we have a computer.

Programming this equation into *Mathematica*, we will be able to find the voltage along the nerve fiber as it starts as a half-sine pulse of 0.5 ms duration with a peak of 60 mV. There is now no wire in the axon, and we will see the voltage response of the nerve in time and space as the signal moves down the axon.

A note of caution is in order. We must be careful to set the initial conditions such that the voltage at t = 0 is zero everywhere, and that the initiating voltage ramps up from there. We must also be careful to set the boundary conditions such that the *maximum* x in the PDE solution term

(D [u [x, t] , x] / . x→ 4) == 0 is greater than the region we are examining.

We will solve the H–H Partial Differential Equation and plot the voltage response at x=0, x=0.5, x=1.0, x=1.5, x=2.0, and x=2.5 cm and from the observed maximum points, we will find the speed of the action potential.

Using the voltage-clamp values of α, β, γ, δ, ϵ, η at T = 6.3 °C, Hodgkin and Huxley obtained a nerve-conduction velocity of 12.4 m/s. The *Mathematica* results on the next page show the nerve conduction velocity at 6.3 °C to be 12.5 m/s.

```
Clear[h]; (*Action potential calculation at T=6.3 °C*)
c = .001; a = .0238; ρ = 35.4;
eq3 = {  a
        ────  D[u[x, t], x, x] ==
        2 * ρ

    c * D[u[x, t], t] + .036 * n[x, t]^4 * (u[x, t] + 12) +
      .120 * m[x, t]^3 * h[x, t] * (u[x, t] - 115) + .0003 * (u[x, t] - 10.6),
    u[x, 0] == 0, u[0, t] == (60 * Sin[2 π * t] * If[0 < t < 0.5, 1, 0]),
    (D[u[x, t], x] /. x → 4) == 0};

eq4 = {D[n[x, t], t] ==   .01 * (10 - u[x, t])
                        ───────────────────────── * (1 - n[x, t]) -
                        Exp[(10 - u[x, t]) / 10] - 1

      .125 * Exp[-u[x, t] / 80] * n[x, t], n[x, 0] == .3177};

eq5 = {D[m[x, t], t] ==   .1 * (25 - u[x, t])
                        ───────────────────────── * (1 - m[x, t]) -
                        Exp[(25 - u[x, t]) / 10] - 1

      4.0 * Exp[-u[x, t] / 18] * m[x, t], m[x, 0] == .0529};

eq6 = {D[h[x, t], t] == .07 * Exp[-u[x, t] / 20] * (1 - h[x, t]) -
                1
        ───────────────────────── * h[x, t], h[x, 0] == .5961};
        Exp[(30 - u[x, t]) / 10] + 1

soll = NDSolve[{eq3, eq4, eq5, eq6},
    {u[x, t], n[x, t], m[x, t], h[x, t]}, {x, 0, 4}, {t, 0, 3.5}];
Plot3D[Evaluate[u[x, t] /. soll[[1]]], {x, 0, 1.5}, {t, 0, 3.5}],
DisplayFunction → $DisplayFunction, ViewPoint → {2.5, -1.5, .95}];
Clear@f; f[x_, t_] = u[x, t] /. soll[[1]]
Pictures = Table[f[x, t], {x, 0, 3.0, .5}]
Plot[Evaluate[Pictures], {t, 0, 3.5}, Frame → True, PlotRange → All]
```

March of the action potentials

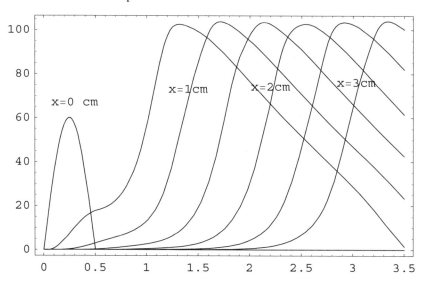

If we evaluate the time at which the maximum is reached at x = 1 cm, it is at t = 1.7 ms, the time of the maximum voltage at x = 3 cm is 3.3 ms. Therefore, at T = 6.3 °C, the speed of the action potential is $v = \frac{\Delta x}{\Delta t} = \frac{2 \text{ cm}}{1.6 \text{ ms}} = 12.5$ m/s.

If we examine a 3-D plot of voltage in time and space, we see the initiating half-sine voltage input on the lower left, followed within a millisecond by the rising hill of the action potential. The action potential then heads off upwards and to the right as it moves down the nerve axon.

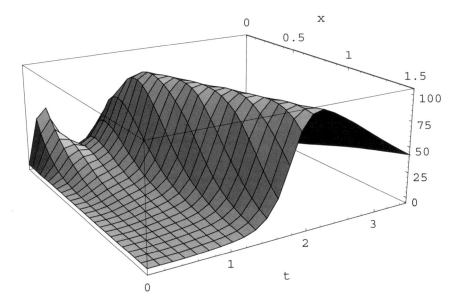

3-D plot of the action potential

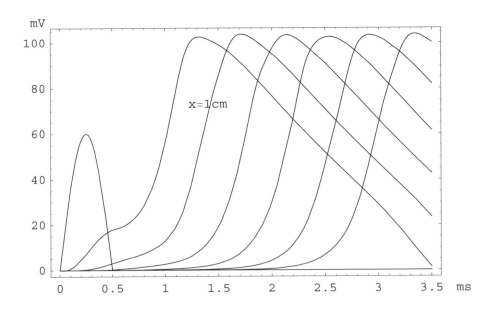

If we take a cross-section of the above 3-D curve, and station ourselves at, say x = 1 cm, then we will see the action potential rise from zero, peak at 100-mV, and then descend almost linearly to zero.

Finally, we may check the conduction velocity v by creating a Table of values of voltage versus time and distance. The maximum voltages (102-103 mV) occur at $(x = 1.0$ cm, $t = 1.7$ ms), $(x = 2.0$ cm, $t = 2.5$ ms), and $(x = 3.0$ cm, $t = 3.3$ ms).

Therefore the conduction velocity is $v = \frac{\Delta x}{\Delta t} = \frac{1 \text{ cm}}{0.8 \text{ ms}} = 12.5$ m/s.

```
TableForm[Transpose@Prepend[
    Transpose@Table[f[x, t], {t, 0, 3.5, .1}, {x, 0, 3.0, 0.5}],
    Table[StyleForm[ToString@t, FontWeight → "Bold"],
      {t, 0, 3.5, .1}]] // Chop,
  TableHeadings -> {None, Prepend[
    Table[StyleForm["x(" <> ToString@x <> ")", FontWeight → "Bold"],
      {x, 0, 3.0, 0.5}], StyleForm["t(ms)", FontWeight → "Bold"]]}]
```

t(ms)	x(0)	x(0.5)	x(1.)	x(1.5)	x(2.)	x(2.5)
0	0	0	0	0	0	0
0.1	35.27	0.689	0.010	-0.00	-0.00	-0.00
0.2	57.03	4.562	0.075	-0.00	-0.00	-0.00
0.3	57.03	10.05	0.553	0.002	-0.00	-0.00
0.4	35.26	14.87	1.575	0.063	-0.00	-0.00
0.5	-0.01	17.80	2.956	0.232	0.004	-0.00
0.6	-0.02	19.27	4.417	0.528	0.029	-0.00
0.7	-0.00	22.48	5.781	0.926	0.083	0.002
0.8	-0.00	28.50	7.332	1.394	0.171	0.011
0.9	-0.00	38.98	9.507	1.941	0.292	0.028
1.	-0.00	56.15	13.00	2.634	0.448	0.055
1.1	-0.00	79.16	19.26	3.617	0.647	0.094
1.2	-0.00	97.41	30.73	5.178	0.913	0.147
1.3	-0.00	**102.5**	48.83	7.804	1.302	0.218
1.4	-0.00	101.5	69.20	12.18	1.916	0.318
1.5	-0.00	99.19	85.72	19.27	2.932	0.465
1.6	-0.00	95.90	98.97	30.68	4.623	0.697
1.7	-0.00	91.65	**103.7**	47.98	7.414	1.071
1.8	-0.00	86.82	102.3	70.25	11.93	1.685
1.9	-0.00	81.76	99.09	87.61	19.10	2.689
2.	-0.00	76.62	95.19	97.66	30.11	4.326
2.1	-0.00	71.47	90.76	**103.0**	45.94	6.958
2.2	-0.00	66.38	85.96	102.6	66.81	11.20
2.3	-0.00	61.40	80.96	99.47	88.52	18.31
2.4	-0.00	56.53	75.88	95.48	99.28	29.91
2.5	-0.00	51.78	70.81	91.04	**102.3**	46.46
2.6	-0.00	47.14	65.81	86.25	102.1	64.84
2.7	-0.00	42.59	60.90	81.26	99.54	84.33
2.8	-0.00	38.06	56.11	76.19	95.81	99.26
2.9	-0.00	33.50	51.44	71.12	91.43	**103.3**
3.	-0.00	28.79	46.87	66.11	86.64	102.4

The Hodgkin–Huxley Equations of Nerve Conduction

After 50 years, the Hodgkin and Huxley equations are still the best descriptor of nerve conduction. This is because the equations were developed using an *empirical model* where the conductivity of the nerve was determined from experimental measurements, without regard to the precise structure of the nerve membrane. One of the first new applications of the H-H theory was to *myelinated nerve,* which is the coated nerve – with gaps – found in the higher animals (humans included !). One of the first computations of nerve velocity in myelinated nerve was done by Richard Fitzhugh in 1962.

If you would like to try and model myelinated nerve conduction, the details are given in R. Fitzhugh *Biophysical Journal 2* , 11-21 "Computation of Impulse Initiation and Conduction in Myelinated Nerve".

For more information on nerve conduction, see Problem #5-4 in the Challenge problems for this chapter, where temperature effects on nerve conduction and impulse initiation are explored.

For the very latest on the molecular structure of nerve membrane and potassium-sodium flow, see recent issues of the *Biophysical Journal.*

5.5   Fusion Reactor

Thermonuclear Fusion

Suppose we are able to magnetically confine a deuterium plasma at a particle density $N = 10^{16}$ deuterons/cm^3 and at temperature $T = 10^8$ Kelvin degrees in a Maxwell-Boltzmann probability distribution. Given that the reaction cross-section for deuterons of kinetic energy E coming from an accelerator and striking stationary deuterons is [§]

$$\sigma(E) = \frac{288}{E} \exp(-45.8/\sqrt{E}) \text{ barns (with E in keV)}$$

We shall first use this result to compute the rate of energy generation per cubic meter from deuterons alone, given that each D + D fusion liberates 3.65 MeV of energy.

Secondly, let us take into account that for every <u>two</u> D + D reactions we obtain <u>one</u> tritium atom which reacts T + D → α + n with an energy release of 17.6 MeV. If all tritium reacts as soon as it is generated, what is the total rate of fusion energy (D + D and T + D) per cubic meter of plasma ?

Let's begin by writing all variables in terms of center-of-mass coordinates.

$$m_* = \frac{m_1 m_2}{m_1 + m_2} = \frac{m_D}{2} = \text{reduced mass}$$

$$E_* = \tfrac{1}{2} m_* v^2 = \frac{E}{2} = \text{relative K.E. of colliding nuclei}$$

The Maxwell-Boltzmann probability distribution is given by

$$f(v)\, dv = \left(\frac{m_*}{2\pi kT}\right)^{3/2} \exp\left(-\frac{m_* v^2}{2kT}\right) 4\pi\, v^2\, dv$$

$$\sigma(E_*) = \frac{288}{2 E_*} \exp(-45.8/\sqrt{2 E_*}) \text{ reaction cross-section in barns (E}_* \text{ in keV)}$$

Now consider a collision of two deuterium nuclei at velocity v. An incoming particle (1) moving at v relative to particle (2) encounters $\overline{v \sigma N_2}$ nuclei in one second. The bar represents an average over the Maxwell-Boltzmann distribution.

The reaction rate is $R = N_1 N_2 (\overline{v\sigma})_{DD} = N_1 N_2 \int_0^\infty (v\,\sigma) f(v)\, dv$

[§] Arnold, Phillips, Sawyer, Stovall, and Tuck (1954) Physical Review <u>93</u>, 483

Since the reacting particles are identical, we must be careful not to count each collision twice

$$R = \frac{N_D^2}{2} \int_0^\infty (v\,\sigma)\left(\frac{m_*}{2\pi kT}\right)^{3/2} \exp\left(-\frac{m_* v^2}{2kT}\right) 4\pi\, v^2\, dv$$

Now, expressing everything in terms of $E_* = \frac{1}{2}m_* v^2$,

$$R = \frac{N_D^2}{2}\left(\frac{2}{m_*\pi}\right)^{1/2} \frac{a}{(kT)^{3/2}} \int_0^\infty e^{-b\,E_*^{-1/2}}\, e^{-E_*/kT}\, dE_*$$

$a = 288$ keV·barns, $b = \frac{45.8}{\sqrt{2}}$ keV$^{1/2}$, $kT = 8.62$ keV, $m_* = \frac{m_D}{2} = 931.5$ MeV/c^2

The above integral cannot be solved analytically. Therefore we will use *Mathematica* to numerically evaluate the integral from $E_* = 1$ to $E_* = 101$ keV

$$\int_1^{101} e^{-b\,E_*^{-1/2}}\, e^{-E_*/kT}\, dE_*$$

```
In[1]:=   α = 8.62; b = 45.8/√2 ;    f[x_] = Exp[-x / α] * Exp[-b/√x];

          NIntegrate[f[x], {x, 1, 101}]
```

```
Out[2]=  0.00278303
```

Then the reaction rate of D + D per cubic cm per second is

$$R_{DD} = \frac{10^{32}}{2}\left(\frac{2*9\times10^{20}}{\pi*931500}\right)^{1/2} \frac{288\times10^{-24}}{(8.62)^{3/2}} (.002783)$$

```
In[3]:=   n = 10^16; (*R = # reactions/cm^3/s*)

          R = n^2/2 ( (2*9*10^20)/(π*931500) )^(1/2) * ( (288*10^-24)/(8.62)^(3/2) ) * .002783
```

```
Out[3]=  R = 3.92722 × 10^13
```

At 3.65 MeV per D + D reaction, $P_{DD} = 23$ MW/ m^3.
If all tritium reacts as soon as it is produced, with $U_{DT} = 17.6$ MeV
and $U_{DD} = 3.65$ MeV, then the total reactor power from both D+D
and D+T is

$$P = R_{DD}\left[U_{DD} + \frac{U_{DT}}{2}\right] = 78\text{ MW}/\text{m}^3.$$

Maxwell–Boltzmann and the Gamow Peak

The Maxwell–Boltzmann distribution tells us that the most probable energy of the colliding deuterium particles is

$$\overline{E} = kT = (1.38\times10^{-23} \text{ J/K} * 10^8 \text{K}) / (1.6\times10^{-16}\text{J/keV}) = 8.62 \text{ keV}.$$

However, this is not the energy at which most D + D reactions take place. To find the most probable energy of reaction, we need to find the maximum of the function inside the reaction integral

$$R_{DD} = \frac{N_D^2}{2}\left(\frac{2}{m_*\,\pi}\right)^{1/2} \frac{a}{(k\,T)^{3/2}} \int_0^\infty e^{-b\,E_*^{-1/2}}\, e^{-E_*/kT}\, dE_*$$

We want to maximize $\qquad g(E_*) = e^{-b\,E_*^{-1/2}}\, e^{-E_*/kT}$

$In[15]:= \quad \alpha = 8.62; \; \beta = \dfrac{45.8}{\sqrt{2}}; \; g[x\_] = Exp[-x / \alpha] * Exp[-\beta\, x^{-1/2}];$

$\qquad Plot[\{Exp[-x / \alpha] * Exp[-\beta\, x^{-1/2}]\}, \{x, 1, 100\}]$

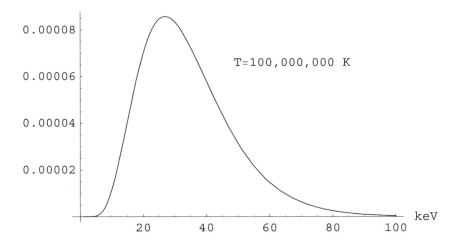

Reaction probability as a function of energy

$In[17]:= \quad Maximize[\{g[x], x > 0\}, x]$

$Out[17]= \quad \{0.0000856866, \{x \to 26.9082\}\}$

This is the so-called "Gamow peak". So even though the average energy of the particles in the deuterium plasma is 8.62 keV, most reactions occur for only the most energetic deuterons, those with three times the average energy. The most probable energy of reaction occurs at 26.9 keV.

This gives us some idea of the difficulty in achieving nuclear fusion. Let us plot the Gamow function at T = 50,000,000 K (which is 3 times the core temperature of the sun) and also at T = 100,000,000 K, to see what effect doubling the temperature has on the reaction probability.

```
In[77]:= α = 8.62; β = 45.8/√2; δ = .5 α;
         Plot[{Exp[-x/α] * Exp[-β x^-1/2], Exp[-x/δ] * Exp[-β x^-1/2]},
         {x, 1, 100}, PlotRange → {0, .0001}, AxesLabel → {"keV", " "},
         Prolog → Text["100,000,000 K", {53, .000066}]]
```

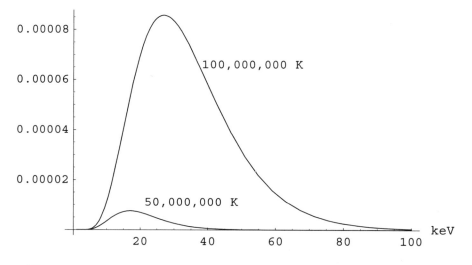

Difficulty of starting a "nuclear fire." The power output at each temperature is proportional to the area under each curve

The power output of our theoretical fusion reactor is only 3 MW/m³ at 50 million degrees, however it increases to 78 MW/m³ at 100 million degrees K. But achieving actual fusion power is far more difficult. The first problem is *Bremsstrahlung* (German for braking-radiation) and this is the effect of electrons being accelerated in the electric field of the deuteron ions. As they are accelerated, the electrons emit radiation (X-rays !) and remove heat from the plasma. A formula for loss of energy due to Bremsstrahlung radiation is

$$P_{Brem} = 4.8 \times 10^{-37} \, N_{ion} N_e \sqrt{T(keV)} \;\; W/m^3$$
(For $N_{ion} = N_e = 10^{22}/m^3$, T = 8.62 keV, P_{Brem} = 140 MW/m³).

This means that we will have to heat a deuterium plasma to a temperature greater than 10^8 K or we will have to start with a deuterium-tritium mix. Otherwise the Bremsstrahlung losses will cool the plasma and the nuclear fire will go out.

Another problem is the reaction products of the deuterium-tritium reaction $D + T \rightarrow \alpha + n$. The α-particle carries about 4 MeV of kinetic energy and is easily captured in the plasma, but the neutron with no electric charge and 14 MeV of energy will escape the reaction chamber. So the problem is not only achieving temperature, but maintaining it, and also shielding against a flood of X-rays and high-energy neutrons from the reactions themselves.

The figure-of-merit for achieving and maintaining fusion reactions is the Triple-Product

N T t = (Particle density # ions/ m³) × (Temperature keV) × (Confinement time sec)

Some of the best results to date in world fusion research are with Tokamak magnetic confinement machines where a huge electric current of approx 7 million amperes is started in deuterium or $D + T$ gas, heating it into the keV range and simultaneously creating a magnetic field which confines the plasma. Best efforts in temperature are in the Japanese Tokamak, JT-60, where T=45 keV and confinement times of t = 28 seconds have been reached however the particle density was only 10^{13} deuterons/cm³.

The next generation of fusion machines are presently being constructed. The ITER (International Tokamak Experimental Reactor) is presently being built in the South of France. It's completion date is in 2015. An inertial confinement device, the NIF (National Ignition Facility) which will use lasers to ignite a plasma will begin operations at Lawrence Livermore in California in 2010.

For more on thermonuclear fusion the reader is referred to the very excellent article by J Rand McNally (Jan 1982) "*Physics of Fusion Fuel Cycles*" Nuclear Technology/Fusion 2 , pp 1-28.

5.6   Space Shuttle Launch

The workhorse of the United States space program from 1980 to 2008 is the Space Shuttle. The Shuttle consists basically of a 40-meter long airplane strapped to a main fuel tank and two solid-fuel booster rockets.

In MKS units, the Equations of Motion of the Space Shuttle are

$$my'' = (T - D) \cos (\epsilon t) - mg_o \left(\frac{R}{R+y} \right)^2 \qquad y_o = 0 \qquad y'_o = 0$$

$$mx'' = (T - D) \sin (\epsilon t) \qquad\qquad\qquad x_o = 0 \qquad x'_o = 0$$

This will take us from launch to booster-rocket separation. Let's look carefully at each term in these equations. First, the mass of the Shuttle, its rockets, and its fuel $m = m_o - \alpha t$. The mass declines as the fuel is burnt, at a rate $\alpha = 9800$ kg/s during the first 120 seconds. The Thrust of the Shuttle, with its solid boosters and main engine is $T = 28.6 \times 10^6$ N. The Drag the shuttle experiences during its lift-off phase is $D = .5\rho AC_D v^2$, where the air density declines with height as $\rho = \rho_o e^{-y/8000}$ with y (the height) measured in meters. The angle the Shuttle makes in-flight with respect to its launch point is $\theta = \text{Tan}^{-1} (x/y)$ and ϵ is the angle the thrust makes with the launch angle, so that the Shuttle will move out along x and allow the booster rockets to fall into the Atlantic Ocean. Last is the gravity term $mg_o \cdot r^{-2}$ where Earth's gravity becomes weaker at great height, however this will only amount to a 2% reduction in g until the Shuttle reaches a height y >50 km.

We can now fill in the details for the first 120 seconds of the Space Shuttle launch. The total mass at lift-off is $m_o = 2.04 \times 10^6$ kg. The initial angle $\theta_o = 0°$ (vertical), and the Shuttle is steered to an angle of approx 50° from the vertical over 120 seconds by allowing the angle from the vertical to increase by $\epsilon = 0.007$ radians every second. We will take the air density $\rho_o = 1.2$ kg/m^3 at the surface, and A and C_D in the drag term to be 100 m^2 and 0.3 respectively.

We are now ready to program these equations into *Mathematica*, and see how the computerized version of the Space Shuttle performs

```
m0 = 2.04 * 10^6; Cd = 0.3; T = 28.6 * 10^6; α = 9800; ε = .007; g = 9.8;
sol = NDSolve[{(m0 - α * t) * y''[t] ==
     (T - 18 * (x'[t]^2 + y'[t]^2) * Exp[-y[t] / 8000]) * Cos[ε * t] -
     (m0 - α * t) * g * (6400 / (6400 + y[t] / 1000))^2, (m0 - α * t) * x''[t] ==
     (T - 18 (x'[t]^2 + y'[t]^2) * Exp[-y[t] / 8000]) * Sin[ε * t],
   x[0] == y[0] == 0, x'[0] == 0, y'[0] == 0}, {x, y}, {t, 0, 120}]
```

```
In[26]:=  InterpFunc1 = x /. sol[[1]]; InterpFunc2 = y /. sol[[1]];
          InterpFunc3 = x' /. sol[[1]]; InterpFunc4 = y' /. sol[[1]];
          tbl = Table[{InterpFunc1[t], InterpFunc2[t]}, {t, 0, 120, 2}];
          ListPlot[tbl,
            Prolog → AbsolutePointSize[4], AspectRatio → 0.65];
```

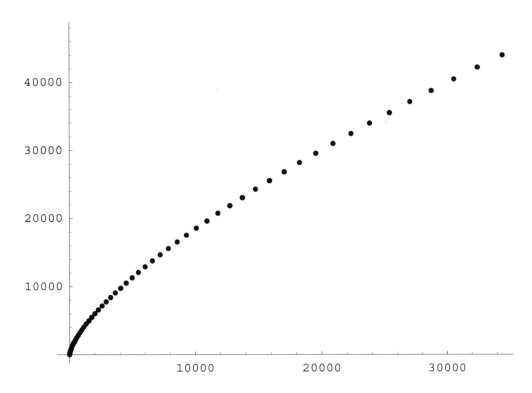

The trajectory of the Space Shuttle from launch to t = 120 seconds, just before booster rocket separation. If we **Table** the results, we will find the height achieved is y = 47 km, and the distance downrange is x = 39 km.

```
In[7]:=  Table[{t, Chop[InterpFunc2[t]], Chop[InterpFunc1[t]],
           Chop[√(InterpFunc3[t]^2 + InterpFunc4[t]^2)],
           Chop[InterpFunc4[t]], Chop[InterpFunc3[t]]},
          {t, 0, 120, 4}] // TableForm
```

The (x,y) coordinates and velocity of the Space Shuttle during its 120-sec ond first-stage flight

t (s)	y (m)	x (m)	v (m / s)	y ' (m / s)	x ' (m / s)
0	0	0	0	0	0
4	34.4714	1.05687	17.4315	17.4134	0.795176
8	140.711	8.53576	36.0239	35.8796	3.22112
12	322.873	29.0812	55.8538	55.3696	7.33858
16	584.989	69.5807	76.9921	75.8506	13.2083
20	930.953	137.164	99.5051	97.2874	20.8907
24	1364.52	239.205	123.458	119.644	30.4468
28	1889.29	383.318	148.915	142.887	41.9378
32	2508.76	577.371	175.946	166.987	55.427
36	3226.3	829.486	204.627	191.922	70.9805
40	4045.23	1148.06	235.044	217.678	88.6692
44	4968.81	1541.79	267.295	244.252	108.571
48	6000.35	2019.69	301.492	271.655	130.772
52	7143.18	2591.16	337.763	299.908	155.369
56	8400.79	3265.98	376.251	329.043	182.471
60	9776.76	4054.43	417.115	359.103	212.202
64	11274.9	4967.29	460.528	390.139	244.7
68	12899.2	6015.92	506.678	422.204	280.118
72	14654.	7212.35	555.762	455.355	318.627
76	16543.6	8569.3	607.989	489.648	360.41
80	18572.8	10100.3	663.576	525.135	405.668
84	20746.3	11819.6	722.746	561.859	454.617
88	23069.3	13742.4	785.731	599.86	507.485
92	25547.	15885.	852.769	639.165	564.519
96	28184.4	18264.5	924.107	679.796	625.98
100	30987.1	20899.1	1000.01	721.766	692.146
104	33960.4	23808.3	1080.74	765.082	763.318
108	37109.6	27012.8	1166.62	809.75	839.824
112	40440.2	30534.5	1257.96	855.772	922.022
116	43957.6	34397.1	1355.14	903.156	1010.31
120	47667.2	38625.7	1458.58	951.913	1105.13

We can visualize the flight of the Shuttle a little better if we Animate the plot, and this is on the Compact Disc as program *SpaceShuttleLaunch*.

For more on the Space Shuttle dynamics, see *The Space Shuttle Operator's Manual* by K.M. Joels, G.P. Kennedy and D. Larkin [Ballantine, 1988]

5.7 Challenge Problems

1. Space Shuttle Phase 2

After solid-booster separation, the mass of the Shuttle and main fuel tank is 690,000 kg. The thrust of the main engine is $T = 5 \times 10^6$ N. If the Space Shuttle main engine burns for 380 seconds with a fuel consumption of 1400 kg/s, find the velocity and height of the Shuttle when the second-phase burn is complete. During the second phase of flight, the Shuttle continues to lean over towards the horizontal, but only at .001 radians per second.

From Section 5.6, the height at booster rocket separation is 47,660 m, and it is 38,620 meters downrange. At the start of the second phase the upward y-velocity is 951 m/s and the x-velocity is 1105 m/s. The heading of the Shuttle is 0.84 radians from vertical at the start of second-phase burn.

During Phase 2 the Shuttle begins to move outward, over the horizon. So be sure when calculating the height to take into account the curvature of the Earth.

2. B–Z Chemical Reaction

The ingredients for a B-Z reaction are the following: 1.415 grams of Sodium bromate, 150 ml 1M H_2SO_4 , 0.175 grams of Cerium-Ammonium-Nitrate, 4.29 grams of malonic acid, all carefully dissolved in distilled water. [See Shakhashiri in the References for the equipment needed and how to safely prepare this solution]. With Ferroin added to the solution, the Cerium ion Ce^{+4} is blue and the Ce^{+3} ion is red. As the reaction proceeds, the solution color changes from green to blue to violet to red, and repeats over and over again as the concentration of the reactants change.

Let $x = [HBrO_2]$, $y = [Br^-]$, and $z = [Ce^{+4}]$ The equations of this B-Z reaction are

$$\epsilon\, x' = x + y - q\, x^2 - xy$$
$$7y' = -y + 2hz - xy$$
$$p\, z' = x - z$$

where $\epsilon = 0.21, p = 14$, $q = 0.006$, $h = 0.75$, $x(0) = 100$, $y(0) = 1$, $z(0) = 10$

Show how the concentration of z varies in time (and therefore, the color change blue-red-blue) and, for good measure, also plot x and y as functions of time. What is the period of oscillation of this reaction? Also try varying the initial concentrations: take $x(0) = 17$, $y(0) = 3$, $z(0) = 5$.

3. Equations of Winemaking

When making wine, the first step is to put the juice from the crushed grapes or grape-juice concentrate into pure water. The amount of natural sugar in the solution may be gauged with a hydrometer. If, for example, our hydrometer reads 1.095 then there are 1140 grams of sugar in a gallon of the grape-juice + pure water mixture.

We now add yeast to the mix. To be scientific we will only add 100 yeast to one gallon. Then place a fermentation lock on the container to allow CO_2 to escape and no foreign material to fall in. Now, here is the winemaking process as seen by the yeast:

1] The yeast break up the sugar into ethyl alcohol + carbon dioxide

$$C_6H_{12}O_6 \longrightarrow 2\ C_2H_5OH + 2\ CO_2 \uparrow$$

2] The yeast extract energy from this process and their numbers grow exponentially at a rate $\quad x' = a\,x \quad$ where $\quad a = 0.125/$ hour

(the yeast number x increases by a factor of e in 8 hours).

3] Because their numbers increase so rapidly, the yeast soon get in each others way.

Therefore the rate of growth of the yeast population really goes as

$$x' = (a - bx)\,x \quad \text{where } b = 10^{-7}a$$

4] All this time, the yeast have been faithfully producing ethanol. The alcohol percentage is increasingly lethal to the yeast at concentrations beyond 4 %, so the full differential equations are

$$\begin{aligned} x' &= (a - bx - cy_*)\,x & x &= \text{\# yeast} \\ y' &= k\,x & y &= \text{alcohol per cent} \\ z' &= -\gamma\,x & z &= \text{remaining sugar (grams)} \end{aligned}$$

initial conditions are $x(0) = 100$, $y(0) = 0$, $z(0) = 1140$ for a one-gallon batch

$$a = 0.125, \quad b = a/10^7, \quad c = a/10, \quad y_* = y - 4 \ (\text{before } y = 4, c = 0)$$

$$\gamma = 10^{11} \text{ sugar molecules/ second/ yeast} = \frac{180*10^{11}}{6*10^{23}} \text{ grams/second/yeast.}$$

$$k = 3.4 \times 10^{-13} \text{ percent/ second/ yeast}$$

How long before the yeast reach their maximum population, x_{max} ?

Solve the above equations of winemaking by tracking x, y, and z for 70 days. What is the final alcohol percentage in the wine at this time ?

4. H–H Nerve Conduction

Alan Hodgkin and Andrew Huxley found that the conductivity of the nerve membrane varies with temperature as $3^{(T-6.3)/10}$. That is, for a nerve fiber at a temperature different than 6.3 °C all the conductivity coefficients $\alpha, \beta, \gamma, \delta, \epsilon, \eta$ are multiplied by the above factor.

Find the conduction velocity of a nerve impulse at T = 18.5 °C by using the H–H nerve conduction model of Section 5.4 with $\alpha, \beta, \gamma, \delta, \epsilon, \eta$ all multiplied by 3.82.

Then use the H–H model to show that when a nerve fiber is at 18.5 °C that a 29-mV, 0.5 ms half-sine pulse will elicit an action potential whereas a 28-mV, 0.5 ms half-sine pulse will <u>not</u> produce an action potential. This is called the "all or nothing response" of a nerve fiber.

5. Flight to the Stars

As a final problem, let's evaluate the possibility of achieving interstellar flight utilizing the most powerful physical processes known.

Assume that matter and antimatter can somehow be stored in a rocket and brought together, producing gamma-ray photons. If the radiation is reflected straight back and the process is 100% efficient, then consider an exploratory trip to Alpha Centauri 4.4 lightyears distant. Accelerate 1/2 way at 1 g, decelerate 1/2 way at 1 g, stay 1 year then accelerate and decelerate at 1 g home.

(A) How long does the trip take as seen by the astronauts on the spaceship, and by the people back home on Earth?

(B) If the total mass of the returning spaceship is 5000 kg (same as Saturn V re-entry vehicle) find the original mass of the starship when it was assembled in space in orbit above the Earth.

Utilize the Impulse–Momentum theorem and the Equations of Special Relativity,

$$m\frac{dv}{d\tau} = -\frac{dm}{d\tau}c = mg \qquad \text{and} \qquad dt = \gamma\, d\tau$$

where Earth-time is t and ship-time is τ and $\gamma = (1 - v^2/c^2)^{-1/2}$

Appendices

Appendix A
Mathematica, Maple, and MatLab

<u>The Mathematical Softwares</u>   *Mathematica*   Maple   MatLab

In comparing the different mathematical softwares, it is instructive to see how each solves the non-linear differential equation

$$\frac{dy}{dt} = y^2 + t^2 \;\; \text{with} \;\; y(0) = 0, \text{from } t = 0 \text{ to } t = 1.$$

Mathematica

```
sol = NDSolve[{y'[t] == y[t]^2+t^2, y[0] == 0}, y, {t, 0, 1}];
Plot[y[t] /. %, {t, 0, 1}]
```

   Maple

```
> dsolve ({diff (y (t), t) = y (t)^2+t^2, y (0) = 0}, y (t), numeric);
> plot (y (t), (t = 0. .1));
```

   MatLab (with Symbolic Math)

```
>> syms y t
>> y = dsolve (' Dy = y (t)^2+t^2', 'y (0) = 0')
>> ezplot (y, [0, 1])
```

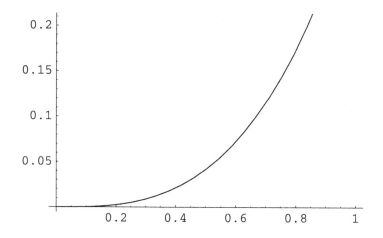

Graphical solution of $y' = y^2 + t^2,$ $y_0 = 0$

Mathematical Softwares   *Mathematica*   Maple   MatLab

As a second example, see how each solves the van der Pol non-linear differential equation

$$y'' + \mu\,(y^2 - 1)\,y' + y = 0 \quad \text{with} \quad \mu = 2,\ y(0) = 1,\ \text{and } y' = 0,$$

from $t = 0$ to $t = 15$.

Mathematica

```
NDSolve[{y''[t] + 2 * (y[t]^2 - 1) * y'[t] + y[t] == 0,
    y'[0] == 0, y[0] == 1}, y, {t, 0, 15}];
Plot[y[t] /. %, {t, 0, 15}]
```

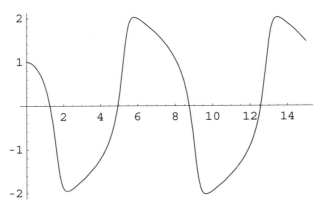

Maple

```
> numsol := dsolve
       ({diff (y (t), t$2) + 2 * (y (t)^2 - 1) * diff (y (t), t) + y (t) = 0,
        D (y) (0) = 0, y (0) = 1}, y (t), numeric);
> with (plots) : odeplot (numsol, [t, (y (t)], 0..15);
```

MatLab (with standard m-file)

First, set $\dfrac{dy1}{dt} = y2$, then $\dfrac{dy2}{dt} = 2 * (1 - y1^2) * y2 - y1$

then write the function-file

```
vanderpol.m
   function yp = vanderpol (t, y)
   yp = [y (2); 2 * (1 - y (1)^2) * y (2) - y (1)];
then call this subroutine with the program
   >> tspan = [0, 15];
   >> y0 = [1, 0];
   >> [t, y] = ode45 ('vanderpol', tspan, yo);
   >> plot (t, y (:, 1))
```

As a third example, consider the predator-prey equations

$$x' = .01x - .001\,xy \quad \text{and} \quad y' = -.015\,y + (.001/40)\,xy \qquad x_o=1000, \quad y_o= 8$$

Mathematica

```
sol = NDSolve[{x'[t] == .01 x[t] - .001 x[t] * y[t],
    y'[t] == -.015 * y[t] + (.001 / 40) * x[t] * y[t],
    x[0] == 1000, y[0] == 8}, {x, y}, {t, 0, 1000}];
p = Plot[Evaluate[{x[t] / 50, y[t]} /. sol,
    {t, 0, 1000}, PlotRange → {0, 22}]];
q = ParametricPlot[{x[t], y[t]} /. sol, {t, 0, 550},
    PlotRange → {{300, 1040}, {4, 19}}]
Show[GraphicsArray[{p, q}]]
```

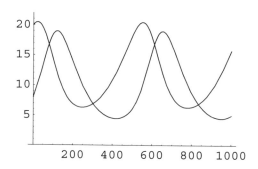 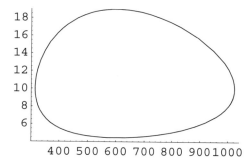

Maple

```
> eq1 := diff (x (t), t) = .01 * x (t) - .001 x (t) * y (t)
> eq2 := diff (y (t), t) = -.015 * x (t) - (.001 / 40) x (t) * y (t)
> sol1 :=
dsolve ({eq1, eq2, x (0) = 1000, y (0) = 8}, {x (t), y (t)}, numeric);
> with (plots) :
> odeplot
    (sol1, [[t, x (t) / 50], [t, y (t)]], 0. .1000, color = [RED, BLUE]);
> odeplot (sol1, [x (t), y (t)], 0. .550, color = [BLACK])
```

MatLab (with standard m-file)

Set y(1) = x, and y(2) = y

$$\frac{dx}{dt} = .01\,x - .001\,xy\,, \quad \frac{dy}{dt} = -.015\,y + (.001/40)\,xy$$

`predprey.m`

```
  function yp = predprey (t, y)
  yp = [.01 * y (1) - .001 * y (1) * y (2);
   -.015 * y (2) + (.001 / 40) * y (1) * y (2)];
```
then call this subroutine with the program
```
>> tspan = [0, 1000];    >> y0 = [1000, 8];
>> [t, y] = ode23 ('predprey', tspan, y0);
>> plot (t, y)        >> plot (y (:, 1), y (:, 2))
```

As a final example, consider **Animations** in *Mathematica,* Maple, and MatLab

In[6]:= *Mathematica*

```
Do[Plot[(Sin[π * x] * Cos[π * t]), {x, 0, 1}, PlotRange → {-1, 1}],
{t, 0, 2, .1}]; (* jump rope *)
```

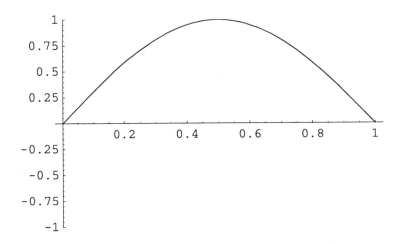

Maple Animation

```
> u := sin (π * x) * cos (π * t)
> with (plots) :
> animate (u (x, t), x = 0. .1, t = 0. .2, frames = 20,
    view = [0. .1, -1..+1], color = BLACK);
```

MatLab Movie (with standard m-file)

to run the program type in

>> movie11

```
% program movie11.m
x = 0 : .1 : 1;
for j = 1 : 200
  t = (j - 1) * .04;
  u = cos (pi * t) * sin (pi * x);
  plot (x, u)
  axis ([0 1 - 1 1])
  pause (0.1);
  M ( :, j) = getframe;
end
```

Cost and Capability comparison for the Mathematical Software (June 2009)

	Universities and Commercial	other full-capability software	Student Edition
Mathematica 6/7	$999	*Mathematica* for classroom $260	$130
Maple 11	$899	Maple for Faculty $180	$125
MatLab 7	$700	(includes PDE toolbox)	$ 99

	Linear and NL ODEs	Stiff DEs	Partial DEs	Animation	Full Graphing Capability
Mathematica 5/6/7	✓	✓	✓	✓	✓
Maple 11	✓	✓	✓	✓	✓
MatLab 7	✓	✓	✓	✓	✓

Note 1. Wolfram Inc also offers Calculation Center 3 for $250. This software may be suitable at the high-school level, however it does not have the full capability of *Mathematica* and is not recommended at the college level.

Note 2: All Student editions are full-capability. However print-outs may include the words "Student Edition" .

Note 3: Many of the mathematical software packages are available for less at www.Academicsuperstore.com, www.CampusTech.com, or www.Ebay.com.

Note 4: *Mathematica* solves PDEs numerically with NDSolve.
Maple solves PDEs with pdsolve.
MatLab utilizes the program pdepe, in the PDE toolbox.

Note 5: Maple may need certain processing limits reset to solve stiff DEs.
MatLab uses the special program ode15s or ode23s to solve stiff DEs
Mathematica's NDSolve automatically tests for stiffness and applies proper techniques.

Note 6: Full graphics capability implies 2-D and 3-D graphs and fitting functions to data.

Note 7: Trial versions of all three mathematical softwares ares are available
Mathematica www.wolfram.com (15 days)
Maple (thru sales rep) www.maplesoft.com
MatLab www.mathworks.com (15 days)

Sources for more information on *Mathematica*, Maple, and Matlab

Websites for *Mathematica*, Maple, and MatLab

Mathematica www.wolfram.com   (see also library.wolfram.com)
Maple   www.maplesoft.com (see also www.mapleapps.com)
MatLab   www.mathworks.com

Books on *Mathematica*, Maple, and MatLab

Mathematica
Heikki Ruskeepaa (2004) *Mathematica* Navigator [Elsevier, 2 ed]
Stephen Wolfram (2003) *Mathematica* Book 5 [Wolfram Publishing]
Martha Abell, James Braselton (2004) Differential Equations with
Mathematica [Academic 3 ed]
Ferdinand Cap (2003) Mathematical Methods in Physics & Engineering
[CRC Press]

Maple
Andre Heck (2003) Introduction to Maple [Springer 3 ed]
Martha Abell, James Braselton (2000) Differential Equations with Maple V
[Academic 3 ed]
Jon Davis (2001) Differential Equations with Maple [Birkhauser]
John Putz (2003) Maple Animation [CRC Press]

MatLab
Duane Hanselman, Bruce Littlefield (2003) Mastering MatLab 7 [PrenticeHall]
Stephen Chapra, Raymond Canale (2002) Numerical Methods for Engineers
[McGraw-Hill 4 ed]
William Palm (2004) Introduction to MatLab 7 for Engineers [Prentice-Hall]
Sergey Lyshevski (2003) Engineering and Scientific Computations using
MatLab [Wiley]
Amos Gilat(2007) MATLAB. An Introduction with Applications [Wiley 3 e]

Appendix B The Compact Disc

The *Mathematica* Compact Disc

Contained on the Compact Disc are all the *Mathematica* program notebooks and Animations in the book. To run any of these programs, it is recommended that they be downloaded to your computer and then run in *Mathematica*.

This way, you may change any file as you wish. However, to re-run any changed file, be sure to press Shift-Enter to re-do all the calculations, otherwise, the old values may still be there.

To run any of the Animations, just double-click on the first of the string of pictures, and you may adjust the run-speed by clicking on the double arrows at the bottom of the page.

Mathematica Animations

`Animation-1.nb`	moving waves and pulses	`(Appendix C)`
`BaseballAnimation.nb`	flight of a baseball	`(Section 1-4)`
`BinaryStarsOrbit.nb`	elliptical orbits of stars	`(Section 3-3)`
`BouncingBall.nb`	follow the bouncing ball	`(Section 1-5)`
`HalleysComet.nb`	elliptical orbit of comet	`(Section 1-2)`
`JupitersMoon.nb`	Ganymede orbiting Jupiter	`(Problem 1-7)`
`LunarLander.nb`	descent to the moon	`(Problem 1-4)`
`NerveConduction.nb`	action potential pulse	`(Section 5-4)`
`SpaceShuttleLaunch.nb`	trajectory of shuttle	`(Section 5-5)`
`VoyageratJupiter.nb`	gravity-assist from Jupiter	`(Section 1-3)`

If you would like a larger Animation, the operational command in *Mathematica* is ImageSize. The largest available is **ImageSize → 6*72.**

See *VoyageratJupiter.nb* or *HalleysComet.nb* for details.

Mathematica Files Chapter 1

`1-1 Motion-of-the-Planets.nb`

`1-2 HalleysNeartheSun.nb`

`1-3 VoyageratJupiter.nb`

`1-4 BaseballwithAirFriction.nb`

`1-5 Bounce.nb`

`1-6 SkyDivingwithParachute.nb`

`Problem 1-1 Satellite`

`Problem 1-2 DragRacer`

`Problem 1-3 SolarSailing`

`Problem 1-4 LunarLander`

`Problem 1-5 MickeyMantle`

`Problem 1-6 SatelliteOrbits`

`Problem 1-7 JupitersMoon`

`Problem 1-8 Golf`

Mathematica Files Chapter 2

2-1 Vibration.nb	Problem 2-1 NYSubway
2-2 ShockAbsorbers.nb	Problem 2-2 LunarSubway
2-3 StepResponse.nb	Problem 2-3 SpeedBump
2-4 RCcircuit.nb	Problem 2-4 RLC-circuit
2-5 VariableStars.nb	Problem 2-5 ImpulseResponse

Mathematica Files Chapter 3

3-1 ModelingwithDiffEqs.nb	Problem 3-1 Pulsar-GenRel
3-2 TrackandField.nb	Problem 3-2 HalleysCometEq
3-3 Pulsar1913+16.nb	Problem 3-3 TheLargestStars
3-4 ExplodingStar.nb	Problem 3-4 AlaskaPipeline
3-5 HeatExchanger.nb	Problem 3-5 HeatExchanger
	Problem 3-6 WomensTrack
	Problem 3-7 BlackHoleSun

Mathematica Files Chapter 4

4-1 PartialDiffEqs.nb	Problem 4-1 RamsauerEffect
4-2 PDE-CopperRod.nb	Problem 4-2 IntermediateStep
4-3 PDE-Heatwave.nb	Problem 4-3 EnergyEigenvalues
4-4 QuantumStep.nb	Problem 4-4 CoolingConcrete
4-4 QuantumBarrier.nb	Problem 4-5 PDE-Steel
4-4 QuantumWell.nb	
4-4 QuantumOscillator.nb	
4-5 HydrogenAtom.nb	
4-6 Deuterium.nb	

Mathematica Files Chapter 5

5-1 LaserPulseDynamics.nb	Problem 5-1 SpaceShuttlePhase2
5-2 B-ZChemicalReaction.nb	Problem 5-2 B-ZReactions
5-3 FoxesandRabbits.nb	Problem 5-3 Winemaking
5-4 PDE-NerveConduction.nb	Problem 5-4 H-HNerveConduction
5-5 FusionReactor.nb	Problem 5-5 FlighttotheStars
5-6 SpaceShuttleLaunch1.nb	

More *Mathematica* Files

* z–1 PDE-Wave.nb	* z–3 AIDSEpidemicUnitedStates.nb
* z–2 PDE-Incropera.nb	* z–4 BinaryStarswithVaporTrails.nb

Appendix C *Mathematica* Commands

The Mathematica Commands. To evaluate any set of *Mathematica* commands one needs to press the keys SHIFT-ENTER on the keyboard.

Basic Functions (note capitalization of functions -- and **square** brackets)

$$\sin(2x) = \textbf{Sin[2x]} \qquad\qquad \cos(x) = \textbf{Cos[x]}$$

$$e^{-x} = \textbf{Exp[-x]} \qquad\qquad \sin^{-1}(x) = \textbf{ArcSin[x]}$$

$$\ln(x) = \textbf{Log[x]} \qquad\qquad \log_{10}(x) = \textbf{Log[10,x]}$$

$$i = \textbf{I} \qquad\qquad \tanh(x) = \textbf{Tanh[x]}$$

Summations (utilize Basic Input Palette)

$$In[9] := \quad \sum_{N=1}^{\infty} \frac{1}{N^2}$$

$$Out[9] = \quad \frac{\pi^2}{6}$$

Complex Numbers

```
In[63]:=  z = 7 + 8 I;
          Re[z]
          Im[z]
          Abs[z]
```

$Out[64] = 7$

$Out[65] = 8$

$Out[66] = \sqrt{113}$

Solving Algebraic Equations (note double equals in equations)

$In[59] := \textbf{Solve[x}^2 + \textbf{14 x} + \textbf{9 == 0, x]}$

$Out[59] = \left\{ \left\{ x \to -7 - 2\sqrt{10} \right\}, \left\{ x \to -7 + 2\sqrt{10} \right\} \right\}$

$In[58] := \textbf{NSolve[x}^3 + \textbf{2 x} + \textbf{3 == 0, x]}$

$Out[58] = \{\{x \to -1.\}, \{x \to 0.5 - 1.65831\,i\}, \{x \to 0.5 + 1.65831\,i\}\}$

Solving Transcendental Equations (note starting value of 1000)

In[60]:= **FindRoot[40 * Log[10, z] + .001 z == 150, {z, 1000}]**

Out[60]= $\{z \to 4372.16\}$

Solving simultaneous equations

In[61]:= **Solve[{x + y == 5, x - y == 1}, {x, y}]**

Out[61]= $\{\{x \to 3, y \to 2\}\}$

CALCULUS

Differentiation

In[51]:=   **D[x², x]**

Out[52]= 2 x

Integration

In[1]:= $\int_0^1 x^2 \, dx$

Out[1]= $\dfrac{1}{3}$

Numerical Integration

In[5]:= α = 8.62; β = 45.8/$\sqrt{2}$;
         f[x_] = Exp[-x / α] * Exp[-β x$^{-1/2}$];
         NIntegrate[f[x], {x, 1, 101}]

Out[7]= 0.00278303

Maximize

In[10]:= α = 8.62; β = 45.8/$\sqrt{2}$;
          Maximize[{Exp[-x / α] * Exp[-β x$^{-1/2}$], x > 0}, x]

Out[11]= $\{0.0000856866, \{x \to 26.9082\}\}$

Minimize

In[12]:= **FindMinimum$\left[400 * \left(\dfrac{1}{x^{12}} - \dfrac{1}{x^6}\right), \{x, 1\}\right]$**

Out[13]= $\{-100., \{x \to 1.12246\}\}$

Differential Equations

In[15]:= DSolve[{y″[x] + 3 y′[x] – 4 y[x] == 0, y[0] == 1, y′[0] == 0},
          y[x], x] // Simplify
          Plot[Evaluate[y[x] /. %], {x, 0, 5}]

Out[15]= $\left\{\left\{y[x] \rightarrow \dfrac{e^{-4x}}{5} + \dfrac{4\, e^{x}}{5}\right\}\right\}$

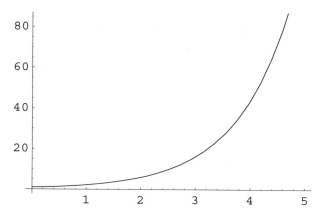

Numerical Solution of any differential equation

In[1]:= sol = NDSolve[{y''[t] + 2 * (y[t]^2 – 1) * y'[t] + y[t] == 0,
          y'[0] == .001, y[0] == 1}, y, {t, 0, 15}];
          Plot[y[t] /. sol, {t, 0, 15}]
          Table[{t, y[t] /. sol, y'[t] /. sol}, {t, 0, 1, .2}] // TableForm

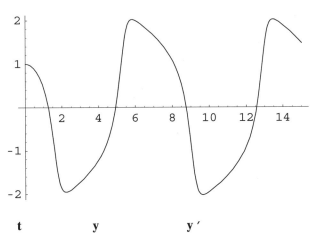

t	y	y′
0	1.	0.001
0.2	0.980234	−0.198464
0.4	0.920465	−0.400748
0.6	0.818515	−0.624963
0.8	0.666473	−0.910267
1.	0.445819	−1.3259

Partial Differential Equations

(*The partial differential equation for heat diffusion*)

$$\frac{\partial u}{\partial t} = \alpha \frac{\partial^2 u}{\partial x^2} \qquad \alpha = .96;$$

Boundary condition   u (x, t) = 0 at x = 0 and x = 20
Initial condition   u (x, 0) = 100 for 0 < x < 20

This PDE is written in *Mathematica* as

```
In[79]:=  α = .96;
          eq4 = {D[u[x, t], t] - .96 * D[u[x, t], x, x] == 0,
             u[x, 0] == Which[x ≤ 0, 0, 0 < x < 20, 100, x ≥ 20, 0],
             u[0, t] == 0, u[20, t] == 0};
          sol4 = NDSolve[eq4, u[x, t], {x, 0, 20}, {t, 0, 40.0}];
          Plot3D[Evaluate[u[x, t] /. sol4[[1]]], {t, 0, 40.0}, {x, 0, 20}],
             PlotRange → All, AxesLabel → {"t", "x", " "}];
```

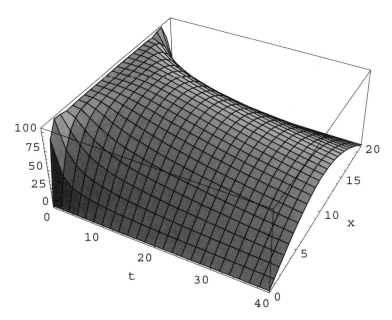

3-D Plot of temperature function u[x, t]

Animation (requires use of a Do-Loop)

In[83]:= **Do[Plot[(Sech[(x - t)])^2, {x, -5, 5}, PlotRange → {0, 1}],
{t, -.5, 8, .5}];**

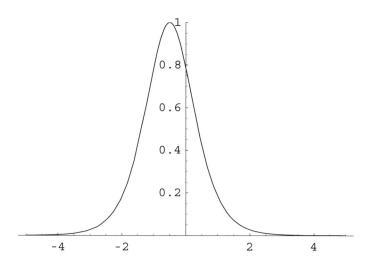

This creates a sequence of pictures, that when scrolled through rapidly,
give the illusion of motion ... Double-click on the plot to see the animation ...

Notes:

1. If you would like to suppress the In/Out labels on calculations, go to the <u>Kernel</u> and turn the **Show In/Out Names** option <u>off</u>

2. There are many more *Mathematica* functions than shown in this brief introduction. See Chapters 1, 3, and 5 for more animations. See Chapters 4 and 5 for more Partial Differential equations. See Chapters 1 thru 5 for many more solutions to linear and non-linear differential equations.

3. For a direct comparison of *Mathematica* commands with Maple and, MatLab, turn to Appendix A.

4. In *Mathematica* 6 and 7, the word **Do** in a Do-Loop is replaced with the word **Animate**.

5. Some changes have been made in *Mathematica* 6 to the "front-end" or user-interface. Thus programs written in *Mathematica* 6 or 7 will not run in *Mathematica* 5, however *Mathematica* 5 programs (with some revision) will run in *Mathematica* 6 and 7. To avoid confusion, please download any programs you wish to run from the Compact Disc, using *Mathematica*-5, *Mathematica*-6, or *Mathematica*-7 files as appropriate for your computer.

Appendix D *MatLab* Commands

MatLab Commands. To evaluate any *MatLab* command, just type the function or m-file name into the command window and press ENTER.

Basic Functions (note no capitalization of functions)

$$\sin(pi/6) \;=\; \mathbf{0.500} \qquad\qquad sind(30) = \mathbf{0.500}$$

$$\exp(-2*x) \;=\; \mathbf{Exp[-2x]} \qquad \text{asin}(x) = \mathbf{sin^{-1}(x)}$$

$$\log(x) \;=\; \mathbf{Ln[x]} \qquad\qquad \log10(x) = \mathbf{Log_{10}[x]}$$

$$i \;=\; i \qquad\qquad\qquad \tanh(x) = \mathbf{Tanh[x]}$$

Summations (syms x k)

```
>> S1 = symsum (1 / k ^ 2, 1, inf)
```

$$S1 \;=\; \frac{\pi^2}{6}$$

Complex Numbers

```
>> z = 7 + 8 * i;
>> real (z)
>> imag (z)
>> abs (z)
```

Out[64]= 7

Out[65]= 8

Out[66]= $\sqrt{113}$

Solving Algebraic Equations (using symbolic math package)

```
>> solve ( ' x² + 14 x + 9 = 0 ')
```

Out[59]= $\left\{ \left\{ x \to -7 - 2\sqrt{10} \right\}, \left\{ x \to -7 + 2\sqrt{10} \right\} \right\}$

```
>> solve ( ' x³ + 2 x + 3 = 0 ')
```

Out[58]= $\{x \to -1.\}, \{x \to 0.5 - 1.65831\,i\}, \{x \to 0.5 + 1.65831\,i\}$

Solving Transcendental Equations (note starting value of 2)

>> **fzero ('tan (.5117 * sqrt (x)) - sqrt (5 - x) / sqrt (x) ', 2}**

{x → 2.4348)

Solving simultaneous equations (syms x y)

>> **solve ('x + y = 5, x - y = 1')**

Out[61]= {{x → 3, y → 2}}

CALCULUS

Differentiation (syms v t)

>> **v = diff ('.252 * exp (-t / 2) ')**

Out[52]= v → .126 * exp (- t / 2)

Integration (syms a x)

>> **int ('2 * a * exp (-x ^ 2) ', 0, inf)**

Out[1]= a * √pi

Numerical Integration

>> **quadl ('exp (-x / 8.62) . * exp [-45.8 . / sqrt (2 * x) ', 1, 101)**

Out[7]= 0.002783

Maximize

>> **fminbnd (' - exp (-x / 8.62) * exp (-45.8 / sqrt (2 * x)) ', 1, 50)**

x → 26.9082

Minimize

>> **[fval x] = fminbnd ('400 * (1 / x ^ 12 - 1 / x ^ 6) ', 1, 2)**

fval → -100, (x → 1.12246)

Differential Equations (syms x y)

```
>> w = dsolve ('D2y + 3 * Dy - 4 * y = 0', 'y (0) = 1', 'Dy (0) = 0', 'x')
>> ezplot (w, [0, 5])
```

Out[15]= $w = \dfrac{e^{-4x}}{5} + \dfrac{4 \, e^{x}}{5}$

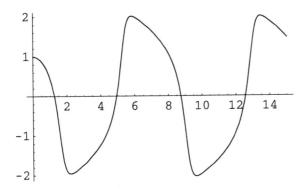

Numerical Solution of any differential equation

```
function AAvdp2
[T Y] = ode45 (@vanderpol2, [0, 15], [1, 0]);
plot (T, Y ( : , 1))
%-------------------
function dydt = vanderpol2 (t, y)
dydt = [y (2); 2 * (1 - y (1) ^ 2) * y (2) - y (1)];
```

t	y	y '
0	1.	0.00
0.2	0.980234	-0.198464
0.4	0.920465	-0.400748
0.6	0.818515	-0.624963
0.8	0.666473	-0.910267
1.	0.445819	-1.3259

Partial Differential Equations

∗ a partial differential equation for heat diffusion ∗

$$\pi^2 \, \frac{\partial u}{\partial t} = \frac{\partial^2 u}{\partial x^2}$$

Boundary condition u (x, t) = 0 at x = 0 and x = 1
Initial condition u (x, 0) = 100 sin (pi ∗ x) for 0 < x < 1

This PDE is written in *MatLab* using pdepe

```
function A41pdeheat
m = 0;
x = linspace (0, 1, 40); t = linspace (0, 1, 10);
sol = pdepe (m, @pdeheatpde, @pdeheatic, @pdeheatbc, x, t);
u = sol ( : , : , 1);
surf (x, t, u)
function [c, f, s] = pdeheatpde (x, t, u, DuDx)
c = pi ^ 2;  f = DuDx;  s = 0;
function u0 = pdeheatic (x)
u0 = 100 ∗ sin (pi ∗ x);
function [pl, ql, pr, qr] = pdeheatbc (xl, ul, xr, ur, t)
pl = ul; ql = 0; pr = ur; qr = 0;
```

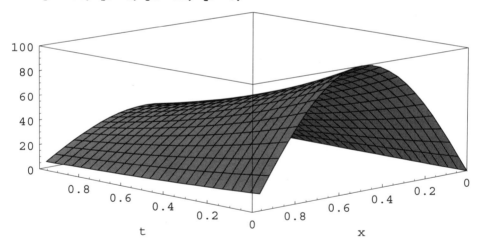

3-D Plot of temperature function **u[x, t]**

Animation (a Movie file in MatLab)

```
% program moviesech.m
x = [-5 : .1 : 5];
for j = 1 : 56
    t = (j - 28) * .1;
    u = sech (x - t) . ^2;
    plot (x, u)
    axis ([-7 7 0 1.4])
    pause (0.1);
    M ( : , j) = getframe;
end
```

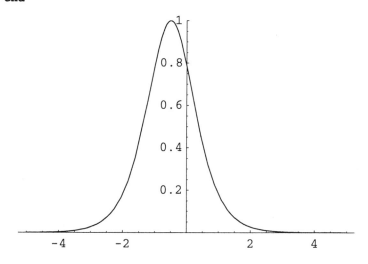

This creates a sequence of pictures, that when scrolled through rapidly,
give the illusion of motion ...   Run the program moviesech to see the animation ...

Notes on MatLab files

1. The MatLab m-files and figures are all written in MatLab-7. Many of the programs (see directory of MatLab programs) require that the symbolic math package be installed. Any programs with Partial Differential Equations require that the PDE-toolbox be installed.

2. When using the various differential equation solvers in MatLab, it is essential to choose a fine enough tolerance for the ode45 and ode23 solvers to succeed. As an example, in A11motionofplanets.m the Relative Tolerance must be set to 1e-4 to close the orbit. Other examples of where the default tolerance of 1e-3 must be changed to 1e-4 are A16, A23, A24, A51, and A54.

3. Some <u>stiff</u> differential equations require the use of ode15s or ode23s. ode15s works well for the vanderPol equation (try AAvdp1000.m) and ode23s works well for Problem 1-7 (pr17) and Problem 5-3 (pr53).

Problem Solutions

Problem1-1 Changing Satellite Speeds

We will start the satellite at the same location as in Section 1.3, except this time, we will give the satellite an initial velocity $v_y = 20$ km/s and $v_x = 0$.

```
M = 1.9 * 10^27; G = 60^2 * 6.67 * 10^-29; (*dist in 10^3 km,time in min*)
sol = NDSolve[{
    x''[t] == M*G* (.013 * 60 t - x[t]) /
        ((x[t] - .013 * 60 t) ^2 + y[t] ^2) ^1.5,
    y''[t] == M*G* (-y[t]) / ((x[t] - .013 * 60 t) ^2 + y[t] ^2) ^1.5,
    x[0] == 10^3 * Cos[40 π / 180], y[0] == -10^3 * Sin[40 π / 180],
    x'[0] == 0, y'[0] == .020 * 60}, {x, y}, {t, 0, 2400}]

InterpFunc1 = x /. sol[[1]]; InterpFunc2 = y /. sol[[1]];
InterpFunc3 = x' /. sol[[1]]; InterpFunc4 = y' /. sol[[1]];
vel[t_] = √ (InterpFunc3[t] ^2 + InterpFunc4[t] ^2) * 1000 / 60;
dist[t_] = √ ((InterpFunc1[t] - .78 t) ^2 + InterpFunc2[t] ^2);
ParametricPlot[{x[t], y[t]} /. sol, {t, 0, 2400},
    Prolog → Circle[{490, 1}, 70], AspectRatio → Automatic];
```

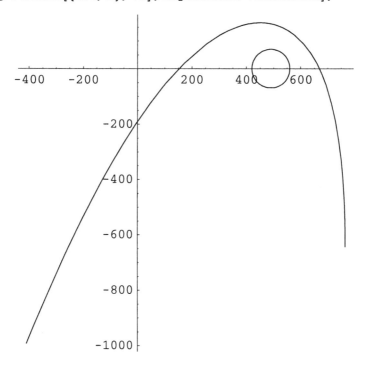

Path of satellite in-front-of Jupiter. All distances in thousands of kilometers. Satellite is slowed from 20 km/s going in to 11 km/s going out.

```
Table[{t, InterpFunc1[t], InterpFunc2[t], vel[t],
    (.78 t), dist[t]}, {t, 0, 2400, 60}] // TableForm
```

t(min)	x(1000 km)	y(1000 km)	v(km/s)	x(Jup)	distance to Jup(1000km)
0	766.0	-642.7	20.	0	1000.
60	765.3	-570.2	20.31	46.8	917.342
120	763.1	-496.4	20.69	93.6	833.544
180	759.0	-421.2	21.15	140.	748.468
240	752.4	-344.4	21.71	187.	661.959
300	742.6	-265.6	22.43	234.	573.874
360	728.3	-184.5	23.37	280.	484.149
420	707.4	-100.8	24.67	327.	393.034
480	675.9	-14.46	26.54	374.	301.933
540	625.8	71.959	29.22	421.	216.925
600	543.7	143.72	31.36	468.	162.461
660	438.4	166.07	28.27	514.	182.776
720	348.3	144.53	23.68	561.	257.624
780	278.3	107.30	20.65	608.	347.046
840	221.6	65.350	18.70	655.	438.477
900	173.4	21.973	17.37	702.	528.972
960	131.2	-21.69	16.40	748.	617.883
1020	93.45	-65.23	15.67	795.	705.169
1080	58.92	-108.4	15.09	842.	790.946
1140	27.01	-151.3	14.62	889.	875.368
1200	-2.77	-193.8	14.23	936.	958.582
1260	-30.8	-236.0	13.90	982.	1040.72
1320	-57.3	-277.8	13.62	1029	1121.9
1380	-82.6	-319.3	13.37	1076	1202.22
1440	-106.	-360.5	13.16	1123	1281.76
1500	-130.	-401.4	12.97	1170	1360.6
1560	-152.	-442.0	12.80	1216	1438.8
1620	-174.	-482.4	12.65	1263	1516.42
1680	-194.	-522.6	12.51	1310	1593.51
1740	-215.	-562.6	12.39	1357	1670.1
1800	-235.	-602.3	12.27	1404	1746.24
1860	-254.	-641.9	12.17	1450	1821.96
1920	-273.	-681.3	12.07	1497	1897.29
1980	-291.	-720.5	11.99	1544	1972.25
2040	-309.	-759.6	11.91	1591	2046.87
2100	-327.	-798.5	11.83	1638	2121.17
2160	-344.	-837.3	11.76	1684	2195.18
2220	-361.	-876.0	11.69	1731	2268.9
2280	-378.	-914.6	11.63	1778	2342.36
2340	-394.	-953.0	11.57	1825	2415.56
2400	-410.	-991.3	11.52	1872	2488.53

Problem 1-2 Drag Racing

Equations of motion for the drag racer

$$v' = \mu g - kv^2 \quad \text{and} \quad x'' = \mu g - kx'^2$$

Terminal velocity v_t is reached when $kv^2 = \mu g$ therefore $k = \dfrac{\mu g}{v_t^2}$

To see the form of the velocity,
take a trial case with $\mu = 2$, $v_{term} = 100$ m/s, $g = 9.8$ m/s^2

```
In[29]:= Clear[x, v, t];
         sol =
          DSolve[{v'[t] == 2 * 9.8 - .00196 * v[t]^2, v[0] == 0}, v[t], t] //
          FullSimplify
         Plot[Evaluate[v[t] /. sol, {t, 0, 10}]];
         v[t] /. sol[[1]] // Chop
```

Out[6]= {{v[t]→100. Tanh[0.196 t]}}

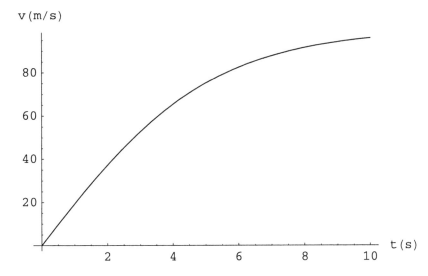

Out[8]= 100. Tanh[0.196 t]

In terms of μ, g, v_t the velocity is $\mathbf{v[t] = v_t\ Tanh\left[\dfrac{\mu\ g}{v_t}\ t\right]}$

To see the form of the distance travelled, solve the equation $x'' = \mu g - kx'^2$
with the same values $\mu = 2$, $v_{term} = 100\,m/s$, $g = 9.8\,m/s^2$, $k = .00196$

```
In[13]:= sol1 = DSolve[{x"[t] == 2 * 9.8 - .00196 x'[t]^2, x[0] == x'[0] == 0},
           x[t], t] // FullSimplify
         Plot[Evaluate[x[t] /. sol1, {t, 0, 10}]];

Out[13]= {x[t] → 510.204 Log[Cosh[0.196 t]]}}
```

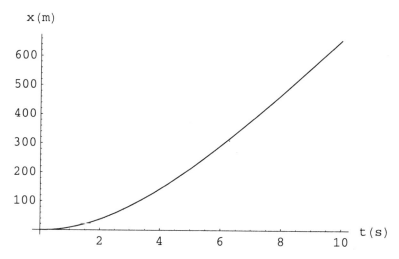

In terms of μ, g, v_t

the distance travelled is $\quad x[t] = \dfrac{v_t^2}{\mu\,g}\,Log\left[Cosh\left[\dfrac{\mu g}{v_t}\,t\right]\right]$

Now we wish to solve the following two equations for μ and v_t

$$v[t] = v_t\,Tanh\left[\dfrac{\mu g}{v_t}\,t\right] \quad \text{and} \quad x[t] = \dfrac{v_t^2}{\mu\,g} * Log\left[Cosh\left[\dfrac{\mu g}{v_t}\,t\right]\right]$$

Since we know $x = 400\,m$, $v = 143\,m/s$ when $t = 4.5\,s$,

$$143 = v_t\,Tanh\left[\dfrac{9.8\,\mu}{v_t} * 4.5\right] \quad \text{and} \quad 400 = \dfrac{v_t^2}{9.8\,\mu}\,Log\left[Cosh\left[\dfrac{9.8\,\mu}{v_t} * 4.5\right]\right]$$

we use *Mathematica* to solve these two equations

```
In[68]:= FindRoot[{143 == v * Tanh[ 44.1 u / v ],

           400 == v² / 9.8 u * Log[Cosh[ 44.1 u / v ]]}, {{u, 2}, {v, 100}}]

Out[68]= {u → 5.20113, v → 160.362}
```

The answer to the required coefficient of friction is a rather surprising $\mu = 5.2$ and the terminal velocity at constant μg acceleration is $v \to 160.36$ m/s.

It would be most interesting to have actual accelerometer data from a high–speed drag race. My guess is the coefficient of friction would exceed 6 at the beginning of the race, and tail off during the race because the tires would not have the same grip on the road.

It is also possible to solve this problem symbolically. Using *Mathematica*, we solve $v' = \mu g - kv^2$ and $x'' = \mu g - kx'^2$

```
In[21]:= Assuming[{Re[μg] > 0, Re[k] > 0},
         DSolve[{v'[t] == μg - k*v[t]^2, v[0] == 0}, v[t], t]]
         Assuming[{Re[μg] > 0, Re[k] > 0},
         DSolve[{x''[t] == μg - k*x'[t]^2, x[0] == x'[0] == 0}, x[t], t]]
```

```
Out[21]=
```

$$\left\{\left\{v[t] \to \frac{\sqrt{\mu g}\ \text{Tanh}\left[\sqrt{k}\ t\ \sqrt{\mu g}\right]}{\sqrt{k}}\right\}\right\}$$

```
Out[22]=
```

$$\left\{\left\{x[t] \to \frac{\text{Log}\left[\text{Cosh}\left[\sqrt{k}\ t\ \sqrt{\mu g}\right]\right]}{k}\right\}\right\}$$

Then, as before, with $v = 143$ m/s and $x = 400$ m, when $t = 4.5$ s,

```
In[24]:= FindRoot[{143 ==
```
$$\frac{\sqrt{\mu g}\ \text{Tanh}\left[\sqrt{k}\ *4.5*\ \sqrt{\mu g}\right]}{\sqrt{k}},$$

```
         400 ==
```
$$\frac{\text{Log}\left[\text{Cosh}\left[\sqrt{k}\ *4.5*\ \sqrt{\mu g}\right]\right]}{k}\}, \{\{\mu g, 2\}, \{k, 1\}\}]$$

```
Out[24]= {μg → 50.9711, k → 0.00198208}
```

Then, $\mu = \dfrac{50.9711}{9.8} = 5.2$ and $v_t = \dfrac{\sqrt{\mu g}}{\sqrt{k}} = 160.36$ m/s

Problem1-3 Solar Sailing

In[5]:= **r = r₀ = 1.5 ∗ 10¹¹ ; (∗distance from sun, 1 AU∗)**

    a = .00592 ;   (∗solar gravity at 1 AU∗)

    g = .00018 ;   (∗solar sail acceleration at 1 AU∗)

At Earth's orbit, $r = r_0$ and we solve for u(x) as a function of sail angle x

In[6]:= $$u[x\_] = \frac{2\,r_0 * (g*Sin[x*\pi/180]*Cos[x*\pi/180]\,{}^\wedge 2)}{r^{.5}\,(a - g*Cos[x*\pi/180]\,{}^\wedge 3)\,{}^\wedge\,.5}\,; \qquad (*\mathbf{EQ\ 1}*)$$

    Plot[u[x], {x, 0, 90}];

    Maximize[{u[x], 0 < x < 90}, x]

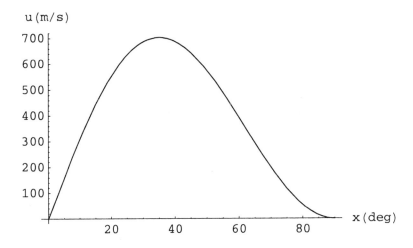

Out[8]= {703.349, {x → 35.0941}}

outward velocity (m / s) vs. sail angle in degrees

Maximum velocity near the Earth's orbit is 703 m/s achieved with a sail angle of 35 degrees. To compute the time to go from Earth's orbit to Mars, notice that u diminishes as $r^{-1/2}$

    b = 703.35 ;   r = 1.5 ∗ 10¹¹ ;

$$T = \int_{r}^{1.5\,r} \frac{z^{.5}}{r^{.5} * b}\,dz$$

    Print["T = ", T / (3.15 ∗ 10^7), " years"]

    T = 3.77837 years

Problem1-4 Lunar Lander

In solving this problem of how to get down to the lunar surface at zero downward velocity, starting from an initial height 10,000 m and an initial downward velocity of 100 m/s, we will adopt the strategy of not turning on the jets for t seconds, and then fire the rockets for T seconds.

Then, $v = v_o + gt$ and $h = 10000 - v_o t - .5 gt^2$

before firing the rockets

and $2.3 T = v_o + gt$ and $\frac{1}{2} 2.3 T^2 = h_o - v_o t - .5 gt^2$

after firing the rockets

By substituting for T in the $\frac{1}{2} 2.3 T^2$ expression, we solve for t.
Using *Mathematica*,

```
v0 = 100; g = 1.7;

NSolve[ .5 * 2.3 * (v0 + g * t)²
                    ─────────── == 10000 - v0 * t - .5 * g * t^2, t]
                       2.3²

{{t → -152.388}, {t → 34.741}}

T = v0 + g * 34.74
    ─────────────
        2.3

T = 69.1557
```

We have now solved for t, the free-fall time, and T the rocket burn time. How much fuel have we expended? F = 2 * 69.15 = 138.3 kg. We are ok. The question now is, do we touch down at zero downward velocity at the surface? Let's plot the entire descent and then Animate the motion. (assume a sideways velocity of $x' = 2$ m/s for visualization)

```
sol = NDSolve[{y'[t] == -(100 + 1.7 t -
        4 * If[t < 34.74, 0, 1] * (t - If[t < 34.74, 0, 34.74])),
      x'[t] == 2, y[0] == 10^4, x[0] == 0}, {x, y}, {t, 0, 104}];
InterpFunc1 = x /. sol[[1]]; InterpFunc2 = y /. sol[[1]];
InterpFunc3 = y' /. sol[[1]];
```

```
ParametricPlot[{x[t], y[t]} /. sol, {t, 0, 103}]
```

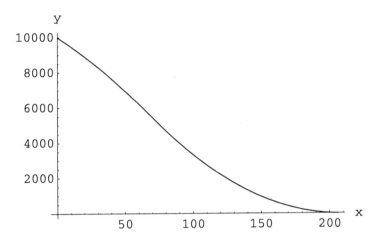

```
Table[{t, InterpFunc1[t], InterpFunc2[t], InterpFunc3[t]},
  {t, 0, 102, 6}] // TableForm
```

t(s)	x(m)	y(m)	↓ v(m/s)
0	0.	10000.	-100.
6	12.	9369.4	-110.2
12	24.	8677.6	-120.4
18	36.	7924.6	-130.6
24	48.	7110.4	-140.8
30	60.	6235.	-151.
36	72.	5301.58	-156.16
42	84.	4406.02	-142.36
48	96.	3593.26	-128.56
54	108.	2863.3	-114.76
60	120.	2216.14	-100.96
66	132.	1651.78	-87.16
72	144.	1170.22	-73.36
78	156.	771.455	-59.56
84	168.	455.495	-45.76
90	180.	222.335	-31.96
96	192.	71.9753	-18.16
102	204.	4.41531	-4.36

We may now Animate the plot

```
Needs["Graphics`MultipleListPlot`"]
Do[MultipleListPlot[
  Table[{InterpFunc1[t], InterpFunc2[t]}, {t, 0, i, 4}],
  SymbolShape → {PlotSymbol[Triangle, 5]}, SymbolStyle → {Hue[2 / 3]},
  PlotRange → {{0, 202}, {-10, 10000}}], {i, 0, 104, 4}]
```

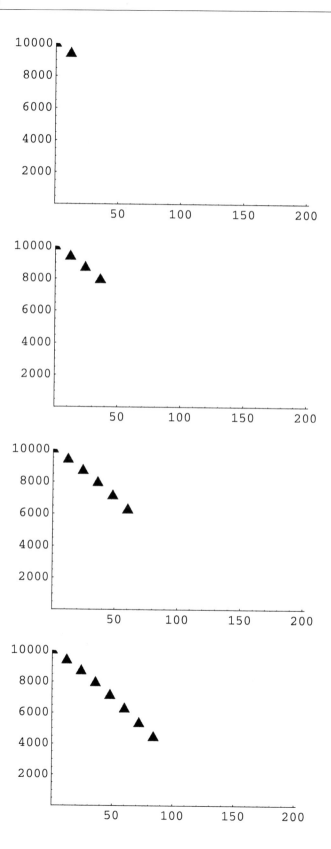

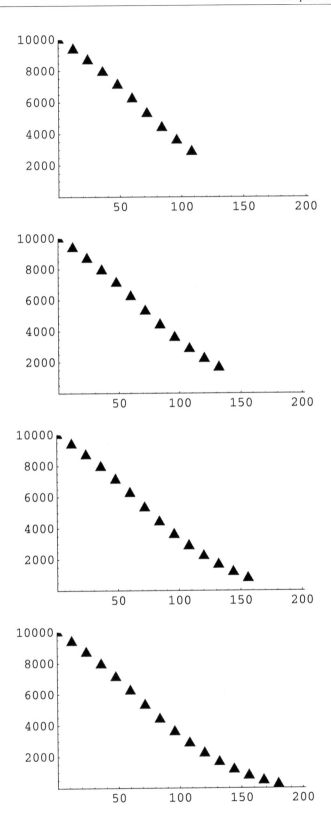

Problem 1-5 Mickey Mantle Home Run

In 1963, Mickey Mantle hit a home run into the third deck at Yankee Stadium. The baseball struck the facade over the third deck some 106 feet above the playing field, and 365 feet from home plate. If the (x,y) coordinates of the ball were (365, 107) then this would have been the first home run ever hit out of Yankee Stadium.

We will assume the same parameters of Section 1-4: a constant drag coeff $C_D = 0.4$, air density $= 0.077$ lb/ft^3, diameter of baseball $= 0.238$ ft, and weight of the baseball mg $= 0.3125$ lb. The drag force is $D = \frac{1}{2} \rho A C_D v^2$

To get the required height of the baseball, we will try an initial angle $\theta = 33°$

$$x'' = -.00219 \, x' \sqrt{x'^2 + y'^2} \qquad x_o = 0 \qquad x'_o = 220 * \cos 33°$$

$$y'' = -32 - .00219 \, y' \sqrt{x'^2 + y'^2} \qquad y_o = 0 \qquad y'_o = 220 * \sin 33°$$

Assuming, as in Section 1-4, that the velocity of the baseball off the bat is 220 ft/s, then the initial x- and y-velocity of the baseball are

```
vx = 220. * Cos[33 * π / 180]          vy = 220. * Sin[33 * π / 180]

    vx = 184.508                           vy = 119.821

sol = NDSolve[{y''[t] == -32 - .00219 y'[t] * √(x'[t]^2 + y'[t]^2),
    x''[t] == -.00219 x'[t] * √(x'[t]^2 + y'[t]^2), x[0] == y[0] == 0,
    x'[0] == 184.5, y'[0] == 120}, {x, y}, {t, 0, 5.4}]

InterpFunc1 = x /. sol[[1]]; InterpFunc2 = y /. sol[[1]];
InterpFunc3 = x' /. sol[[1]]; InterpFunc4 = y' /. sol[[1]];
tbl = Table[{InterpFunc1[t], InterpFunc2[t]}, {t, 0, 5.4, .2}];
ListPlot[tbl, Prolog → AbsolutePointSize[4], AspectRatio → 0.3];
```

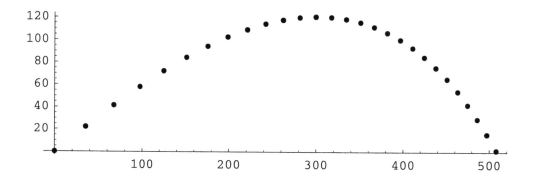

If we **Table** the results, we will find the distance the baseball travelled and the time. From the Table, the baseball passed above the point (365, 107) at $t = 3.2$ seconds.

$$\texttt{Table}\big[\{\texttt{t, Chop[InterpFunc1[t]], Chop[InterpFunc2[t]],}$$
$$\sqrt{\texttt{InterpFunc3[t]}\,\texttt{^2 + InterpFunc4[t]^2}}\,\},$$
$$\texttt{\{t, 0, 5.4, .2\}}\big]\;\texttt{// TableForm}$$

The (x,y) coordinates and velocity of the baseball during its 5.4-second flight

t (s)	x (ft)	y (ft)	v (ft / s)
0	0	0	220.091
0.2	35.2363	22.297	197.612
0.4	67.5446	41.5133	178.854
0.6	97.4139	58.0463	162.984
0.8	125.222	72.2017	149.413
1.	151.266	84.219	137.718
1.2	175.783	94.2888	127.591
1.4	198.966	102.565	118.802
1.6	220.973	109.173	111.179
1.8	241.932	114.217	104.592
2.	261.952	117.783	98.9408
2.2	281.122	119.944	94.148
2.4	299.517	120.764	90.1501
2.6	317.2	120.297	86.8926
2.8	334.221	118.591	84.3258
3.	350.625	115.69	82.4007
3.2	366.447	111.634	81.0673
3.4	381.717	106.463	80.2728
3.6	396.459	100.212	79.9619
3.8	410.693	92.9191	80.0774
4.	424.436	84.6197	80.5613
4.2	437.703	75.3504	81.3567
4.4	450.506	65.1483	82.4089
4.6	462.854	54.0509	83.6668
4.8	474.758	42.0962	85.0836
5.	486.227	29.3225	86.6174
5.2	497.268	15.7685	88.2312
5.4	507.891	1.47279	89.8934

Mantle's prodigious drive would have travelled somewhere around 508 feet.

For more on baseball home runs, see Dan Valenti *Clout! The Top Home Runs in Baseball History* [Stephen Greene Publisher, 1989]

Problem 1-6 Satellite Orbits

Let us first find the velocity of the satellite in its original circular orbit

```
r = 6.6 * 10^6; M = 6 * 10^24;  G = 6.67 * 10^-11;
v0 = NSolve[v² / r == M * G1 / r^2, v]
```

$\{\{v \to -7786.94\}, \{v \to 7786.94\}\}$

Then we set up the Newtonian equations for the new orbit

$$\ddot{x} = -\frac{GM\,x}{(X^2 + Y^2)^{3/2}} \qquad x_o = 6600 \text{ km} \qquad \dot{x}_o = 0$$

$$\ddot{y} = -\frac{GM\,y}{(X^2 + Y^2)^{3/2}} \qquad y_o = 0 \qquad \dot{y}_o = 1.08 * 7.787 \text{ km/s}$$

```
M = 6 * 10^24; G = (60^2) * 6.67 * 10^-20; (*dist in km, time in min*)
sol = NDSolve[{x''[t] == M * G * (-x[t]) / (x[t]^2 + y[t]^2)^1.5,
    y''[t] == M * G * (-y[t]) / (x[t]^2 + y[t]^2)^1.5,
    x[0] == 6.6 * 10^3, y[0] == 0,
    x'[0] == 0, y'[0] == 1.08 * 7.787 * 60}, {x, y}, {t, 0, 1000}]
ParametricPlot[{x[t], y[t]} /. sol, {t, 0, 800}];
```

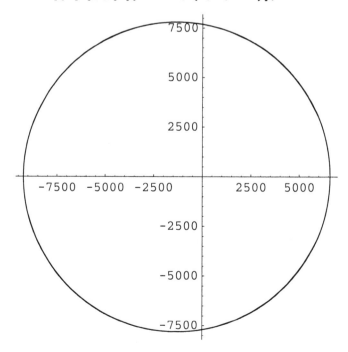

New orbit of the satellite, after rocket burn. Perigee remains at 6600 km from Earth center.

New apogee is at 9230 km from Earth center. (See **Table** for values of distance and velocity)

```
InterpFunc1 = x /. sol[[1]]; InterpFunc2 = y /. sol[[1]];
InterpFunc3 = x' /. sol[[1]]; InterpFunc4 = y' /. sol[[1]];
Rad[z_] = √(InterpFunc1[z]^2 + InterpFunc2[z]^2);
Dist[z_] = √(InterpFunc1[z]^2 + InterpFunc2[z]^2) - 6400;
Vel[z_] := √(InterpFunc3[z]^2 + InterpFunc4[z]^2) / 60;
Table[{t, Rad[t], Dist[t], Vel[t]}, {t, 0, 150, 5}] // TableForm
```

t(min)	R(km)	D(km ↑ E)	v(km/s)
0	6600.	200.	**8.40996**
5	6667.74	267.744	8.33638
10	6859.09	459.086	8.13306
15	7143.51	743.507	7.84223
20	7482.99	1082.99	7.51115
25	7841.59	1441.59	7.17815
30	8190.14	1790.14	6.86891
35	8506.99	2106.99	6.59864
40	8776.94	2376.94	6.37559
45	8989.74	2589.74	6.204
50	9138.72	2738.72	6.08589
55	9219.9	2819.9	6.0222
60	9231.29	2831.29	6.0133
65	9172.64	2772.64	6.05923
70	9045.33	2645.33	6.15973
75	8852.67	2452.67	6.31412
80	8600.35	2200.35	6.52079
85	8297.33	1897.33	6.77639
90	7957.04	1557.04	7.07424
95	7598.87	1198.87	7.40177
100	7249.52	849.522	7.73705
105	6943.01	543.009	8.04588
110	6717.51	317.506	8.28288
115	6607.24	207.24	**8.40206**
120	6631.16	231.163	8.37601
125	6785.03	385.03	8.21099
130	7043.27	643.275	7.94324
135	7369.07	969.065	7.62043
140	7725.03	1325.03	7.28464
145	8079.57	1679.57	6.96559
150	8408.66	2008.66	6.68149

Notice that the period of the new orbit (when v = 8.4 km/s again) is 115 minutes.

Problem 1-7 Jupiter's Moon

Let us write the equations of motion for Ganymede orbiting Jupiter.

First, find the orbital velocity of Ganymede, in a circular orbit,

$$\frac{mv^2}{r} = \frac{mGM}{r^2} \qquad v = 11.26 \text{ km/s}$$

Then program the Equations of motion for Ganymede orbiting Jupiter

$$\ddot{x} = -\frac{GM (x - 13\ t)}{\left((x - 13\ t)^2 + y^2 \right)^{3/2}} \qquad x_o = 0 \qquad \dot{x}_o = 24.26 \text{ km/s}$$

$$\ddot{y} = -\frac{GM\ y}{\left((x - 13\ t)^2 + y^2 \right)^{3/2}} \qquad y_o = -10^6 \text{ km} \qquad \dot{y}_o = 0$$

This takes into account the motion of Jupiter, which is moving along $+x$ at 13 km/s.

```
In[39]:=  G = 6.673 * 10⁻¹¹; M = 1.9 * 10²⁷; r = 10⁹;
          Solve[v² / r == G M / r², v]

Out[41]=  {{v → -11260.}, {v → 11260.}}

M = 1.9 * 10²⁷; G = (60^2) * 6.673 * 10⁻²⁹; (*dist in 10³ km, time in min*)
sol = NDSolve[{
    x''[t] == G * M * (.013 * 60 t - x[t]) / ((x[t] - .013 * 60 t) ^2 + y[t] ^2) ^1.5,
    y''[t] == G * M * (-y[t]) / ((x[t] - .013 * 60 t) ^2 + y[t] ^2) ^1.5,
    x[0] == 0, y[0] == -10^3, x'[0] == .02426 * 60, y'[0] == 0},
    {x, y}, {t, 0, 24000}]

InterpFunc1 = x /. sol[[1]]; InterpFunc2 = y /. sol[[1]];
InterpFunc3 = x' /. sol[[1]]; InterpFunc4 = y' /. sol[[1]];
Table[{t, Chop[InterpFunc1[t]], InterpFunc2[t], (.013 * 60 * t),
    √(InterpFunc3[t] - .78) ^2 + InterpFunc4[t] ^2 * 1000 / 60,
    √(InterpFunc1[t] - .78 t) ^2 + InterpFunc2[t] ^2},
    {t, 0, 10000, 400}] // TableForm
ParametricPlot[{x[t], y[t]} /. sol, {t, 0, 23000} ];
```

Table of values with Time in minutes, Distance in thousands of kilometers

t(min)	x	y	v(km/s)	x(Jup)	dist from Jup (1000 km)
0	0	-1000.	11.26	0	1000.
400	578.96	-963.7	11.26	312.	1000.
800	1138.5	-857.4	11.26	624.	1000.
1200	1660.7	-688.9	11.26	936.	1000.
1600	2130.4	-470.4	11.26	1248.	1000.
2000	2535.9	-217.8	11.26	1560.	1000.
2400	2870.7	50.616	11.26	1872.	1000.
2800	3132.9	315.39	11.26	2184.	1000.
3200	3326.3	557.28	11.26	2496.	1000.
3600	3459.4	758.72	11.26	2808.	1000.
4000	3545.2	905.09	11.26	3120.	1000.
4400	3600.1	985.76	11.26	3432.	1000.
4800	3642.9	994.88	11.26	3744.	1000.
5200	3692.9	931.79	11.26	4056.	1000.
5600	3769.4	801.07	11.26	4368.	1000.
6000	3889.2	612.20	11.26	4680.	1000.
6400	4066.5	378.89	11.26	4992.	1000.
6800	4310.9	118.08	11.26	5304.	1000.
7200	4627.5	-151.2	11.26	5616.	1000.
7600	5015.7	-409.6	11.26	5928.	1000.
8000	5470.2	-638.3	11.26	6240.	1000.
8400	5980.6	-820.6	11.26	6552.	1000.
8800	6532.4	-943.4	11.26	6864.	1000.
9200	7108.3	-997.7	11.26	7176.	1000.
9600	7689.1	-979.5	11.26	7488.	1000.

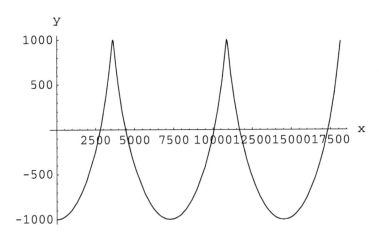

The x-y coordinates of Jupiter's moon. Believe it or not, this is the exact path of the moon as Jupiter moves along + x at 13 km/s. To see how this results in a circular orbit, **Animate** the plot. To see the animation, access *Jupiter'sMoon* on the CD.

Do Loop

```
Needs["Graphics`MultipleListPlot`"]
Do[
  MultipleListPlot[Table[{InterpFunc1[t], InterpFunc2[t]}, {t, i, i}],
   {{0 + .78 * t, 10}} /. {t → i}, PlotRange → {{0, 7600}, {-1100, 1100}},
   SymbolShape → {MakeSymbol[Circle[{0, 0}, 50]],
     MakeSymbol[Circle[{0, 0}, 140]]},
   AspectRatio → Automatic, Ticks → None], {i, 0, 9600, 640}]
```

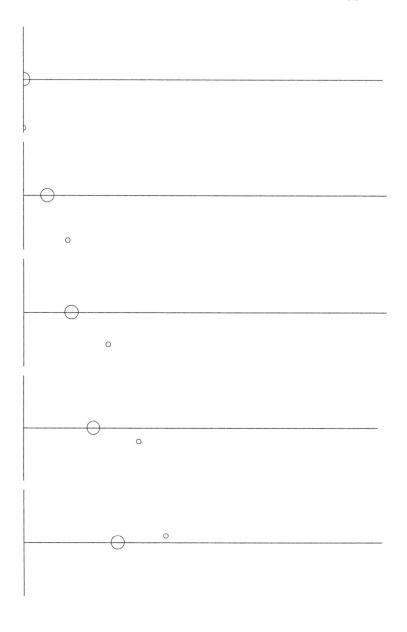

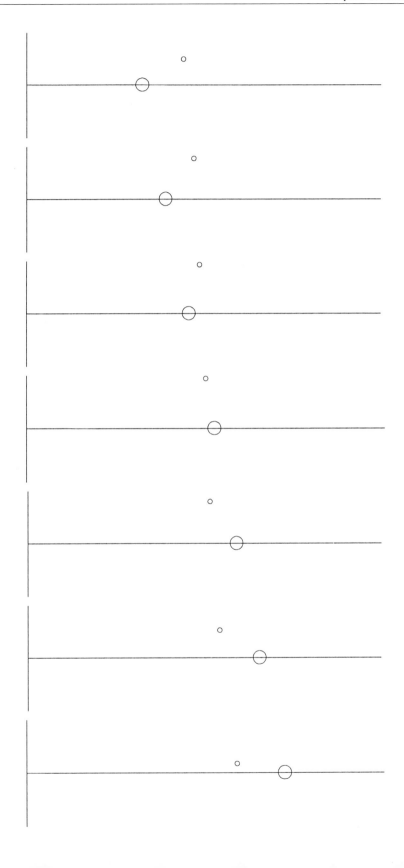

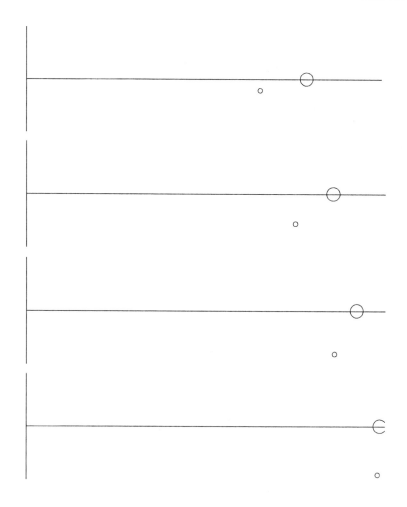

To slow down the animation, just click on the downward-facing chevron. Or, you can change the Cell option "AnimationDisplayTime." Select the cell bracket that is grouping all 6 cells that contains the graphs, go to Format > Options Inspector, then search for "AnimationDisplayTime."

Problem 1-8   Golf

For a golf ball driven off the tee at $10°$ at $v_o = 220$ ft/s, with backspin of 3000 rpm, the Equations of Motion are

$$x'' = -.00188\, x'\, \sqrt{x'^2 + y'^2}\ -.235\, y' \qquad\qquad x_o = 0 \quad x'_o = 220 * \text{Cos } 10$$
$$y'' = -32.16 - .00188\, y'\, \sqrt{x'^2 + y'^2}\ +.235\, x' \quad y_o = 0 \quad y'_o = 220 * \text{Sin } 10$$

Mathematica easily solves these coupled non-linear equations

```
In[60]:=  sol = NDSolve[{y''[t] ==
              -32.16 - .00188 y'[t] * Sqrt[x'[t]^2 + y'[t]^2] + .235 x'[t],
            x''[t] == -.00188 x'[t] * Sqrt[x'[t]^2 + y'[t]^2] - .235 y'[t],
            x[0] == y[0] == 0, x'[0] == 216, y'[0] == 38}, {x, y}, {t, 0, 7.6}]
```

```
In[61]:=  InterpFunc1 = x /. sol[[1]]; InterpFunc2 = y /. sol[[1]];
          InterpFunc3 = x' /. sol[[1]]; InterpFunc4 = y' /. sol[[1]];
          tbl = Table[{InterpFunc1[t], InterpFunc2[t]}, {t, 0, 7.4, .2}];
          v220 =
          ListPlot[tbl, Prolog → AbsolutePointSize[4], AspectRatio → 0.25]
```

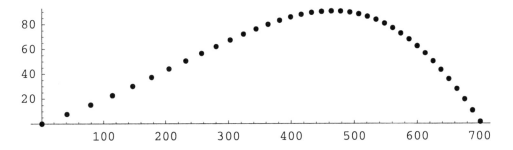

The flight of the golf ball with lift and drag,
initial vel 220 ft / s, spin 3000 rpm, initial angle 10 °

With the lift force included, a golfball launched off the tee at $10°$ and 220 ft/s will carry approximately 700 ft. If we **Table** the results, we will find the distance and height of the drive.

```
In[92]:= Table[{t, Chop[InterpFunc1[t]], Chop[InterpFunc2[t]],
           √InterpFunc3[t]^2 + InterpFunc4[t]^2}, {t, 0, 7.4, .2}] //
         TableForm
```

t (s)	x (ft)	y (ft)	v (ft / s)
0	0	0	219.317
0.2	41.3392	7.6368	201.525
0.4	79.311	15.2693	186.149
0.6	114.368	22.7982	172.706
0.8	146.879	30.1435	160.841
1.	177.151	37.2393	150.282
1.2	205.441	44.0309	140.821
1.4	231.967	50.4725	132.297
1.6	256.919	56.5256	124.582
1.8	280.46	62.1571	117.577
2.	302.733	67.3389	111.203
2.2	323.866	72.0467	105.397
2.4	343.971	76.2598	100.111
2.6	363.151	79.9601	95.3052
2.8	381.497	83.1321	90.9509
3.	399.091	85.7626	87.0248
3.2	416.01	87.8402	83.5096
3.4	432.323	89.3556	80.3918
3.6	448.093	90.3009	77.6611
3.8	463.378	90.6702	75.3084
4.	478.233	90.4587	73.3254
4.2	492.707	89.6638	71.703
4.4	506.847	88.2839	70.4306
4.6	520.694	86.3195	69.4953
4.8	534.289	83.7723	68.8813
5.	547.667	80.6459	68.5702
5.2	560.864	76.9456	68.5407
5.4	573.91	72.6782	68.769
5.6	586.833	67.8522	69.2296
5.8	599.661	62.4777	69.8961
6.	612.419	56.5663	70.7415
6.2	625.127	50.1312	71.739
6.4	637.808	43.1868	72.8631
6.6	650.48	35.7488	74.0892
6.8	663.159	27.8341	75.3946
7.	675.861	19.4603	76.7585
7.2	688.6	10.6461	78.162
7.4	701.389	1.41064	79.5883

For v_o = 200 ft / s

In[1]:= **sol = NDSolve$\Big[$**

$$\Big\{y''[t] == 32.16 - .00188\,y'[t] * \sqrt{x'[t]\,^2 + y'[t]\,^2} + .235\,x'[t],$$

$$x''[t] == -.00188\,x'[t] * \sqrt{x'[t]\,^2 + y'[t]\,^2} - .235\,y'[t],$$

$$x[0] == y[0] == 0,\ x'[0] == 197,\ y'[0] == 34.7\Big\},\ \{x, y\},\ \{t, 0, 8.0\}\Big]$$

In[2]:= **InterpFunc1 = x /. sol[[1]]; InterpFunc2 = y /. sol[[1]];**
     InterpFunc3 = x' /. sol[[1]]; InterpFunc4 = y' /. sol[[1]];
     tbl = Table[{InterpFunc1[t], InterpFunc2[t]}, {t, 0, 6.8, .2}];
     v200 =
     ListPlot[tbl, Prolog → AbsolutePointSize[4], AspectRatio → 0.25]

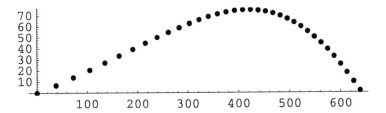

In[6]:= **Table$\Big[\{$t, Chop[InterpFunc1[t]], Chop[InterpFunc2[t]],**
     $$\sqrt{\mathtt{InterpFunc3[t]\,^2 + InterpFunc4[t]\,^2}}\,\},$$
     {t, 0, 6.8, .4}$\Big]$ // TableForm

t (s)	x (ft)	y (ft)	v (ft / s)
0	0	0	200.033
0.4	72.8045	13.828	171.766
0.8	135.547	27.0969	149.742
1.2	190.448	39.2911	132.067
1.6	239.114	50.0491	117.603
2.	282.757	59.1094	105.643
2.4	322.325	66.279	95.7458
2.8	358.581	71.4146	87.6389
3.2	392.15	74.4111	81.1575
3.6	423.555	75.1946	76.1994
4.	453.241	73.7191	72.6867
4.4	481.586	69.9651	70.5331
4.8	508.912	63.939	69.6209
5.2	535.499	55.674	69.7931
5.6	561.583	45.2291	70.8604
6.	587.366	32.6887	72.6193
6.4	613.018	18.1594	74.8704
6.8	638.679	1.76729	77.4341

For $v_0 = 240$ ft / s

$In[7]:=$ **sol = NDSolve$\Big[$**

$\Big\{$**y''[t] == -32.16 - .00188 y'[t] $*$ $\sqrt{x'[t]\wedge2+y'[t]\wedge2}$ + .235 x'[t]**

x''[t] == -.00188 x'[t] $*$ $\sqrt{x'[t]\wedge2+y'[t]\wedge2}$ - .235 y'[t],

x[0] == y[0] == 0, x'[0] == 236, y'[0] == 41.6$\Big\}$, {x, y}, {t, 0, 8.4}$\Big]$

$In[8]:=$ **InterpFunc1 = x /. sol[[1]]; InterpFunc2 = y /. sol[[1]];**
InterpFunc3 = x' /. sol[[1]]; InterpFunc4 = y' /. sol[[1]];
tbl = Table[{InterpFunc1[t], InterpFunc2[t]}, {t, 0, 8.0, .2}];
v240 =
ListPlot[tbl, Prolog $\to$ AbsolutePointSize[4], AspectRatio $\to$ 0.25]

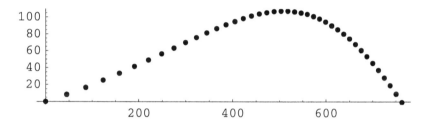

$In[12]:=$ **Table$\Big[\big\{$t, Chop[InterpFunc1[t]], Chop[InterpFunc2[t]],**
$\sqrt{\text{InterpFunc3}[t]\wedge2 + \text{InterpFunc4}[t]\wedge2}\big\}$,
{t, 0, 8.0, .4}$\Big]$ // TableForm

t (s)	x (ft)	y (ft)	v (ft / s)
0	0	0	239.638
0.4	86.0706	16.8119	200.921
0.8	158.537	33.356	172.033
1.2	220.746	48.9679	149.533
1.6	274.984	63.2031	131.467
2.	322.9	75.7506	116.664
2.4	365.736	86.3859	104.401
2.8	404.462	94.9438	94.2294
3.2	439.856	101.302	85.8765
3.6	472.557	105.37	79.1855
4.	503.101	107.087	74.0667
4.4	531.94	106.416	70.4563
4.8	559.461	103.343	68.2793
5.2	585.994	97.8826	67.4214
5.6	611.823	90.0722	67.7188
6.	637.189	79.9776	68.9671
6.4	662.3	67.6902	70.9431
6.8	687.327	53.3249	73.4289
7.2	712.412	37.0168	76.2288
7.6	737.674	18.9164	79.1792
8.	763.203	-0.815301	82.1507

(A)   Golf ball carry (in-air flight) at v_o =200, 220, and 240 ft/s

In[70]:=   **Show[v200, v220, v240]**

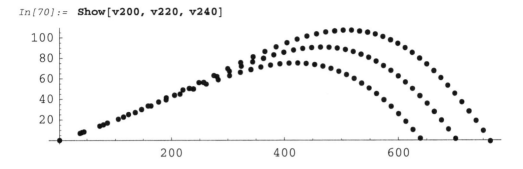

(B)   For an estimate of a Tiger Woods drive, with launch speed of 260 ft/.s and initial angle 10° (Wood's driver speed has been measured at 135 mph.)

In[2]:=   **sol = NDSolve$\Big[\big\{$y''[t] ==**

         -32.16 - .00188 y'[t] $*\sqrt{\text{x'[t]}\,^\wedge 2 + \text{y'[t]}\,^\wedge 2}$ + .235 x'[t],

         x''[t] == -.00188 x'[t] $*\sqrt{\text{x'[t]}\,^\wedge 2 + \text{y'[t]}\,^\wedge 2}$ - .235 y'[t],

         x[0] == y[0] == 0, x'[0] == 256, y'[0] == 45.1$\big\}$, {x, y}, {t, 0, 9.0}$\Big]$

In[3]:=   **InterpFunc1 = x /. sol[[1]]; InterpFunc2 = y /. sol[[1]];**
         InterpFunc3 = x' /. sol[[1]]; InterpFunc4 = y' /. sol[[1]];
         tbl = Table[{InterpFunc1[t], InterpFunc2[t]}, {t, 0, 8.4, .2}];
         v260 =
         ListPlot[tbl, Prolog → AbsolutePointSize[4], AspectRatio → 0.25]

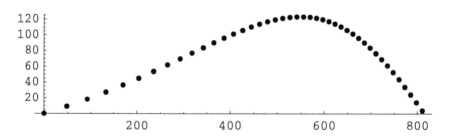

The flight of the golf ball with lift and drag,
initial vel 260 ft / s, spin 3000 rpm, initial angle 10 °

```
In[7]:= Table[{t, Chop[InterpFunc1[t]], Chop[InterpFunc2[t]],
          √InterpFunc3[t]^2 + InterpFunc4[t]^2 },
        {t, 0, 8.4, .4}] // TableForm
```

t (s)	x (ft)	y (ft)	v (ft / s)
0	0	0	259.942
0.4	92.7457	18.297	215.313
0.8	169.944	36.4214	182.745
1.2	235.622	53.6434	157.771
1.6	292.453	69.486	137.931
2.	342.325	83.6211	121.778
2.4	386.636	95.8148	108.429
2.8	426.462	105.896	97.3343
3.2	462.659	113.738	88.1618
3.6	495.922	119.248	80.7174
4.	526.828	122.36	74.8955
4.4	555.867	123.032	70.6345
4.8	583.453	121.245	67.8751
5.2	609.941	117.005	66.5251
5.6	635.635	110.342	66.4388
6.	660.796	101.316	67.4194
6.4	685.645	90.0091	69.2387
6.8	710.37	76.5323	71.6636
7.2	735.125	61.0172	74.4793
7.6	760.037	43.6136	77.5023
8.	785.209	24.4837	80.5853
8.4	810.718	3.79753	83.6162

Depending on the run of the ball after its in-air flight, Tiger Woods can consistently drive the ball between 270 and 290 yards.

Problem 2-1 New York to Leningrad

A famous physics problem is to consider travel by gravity by tunneling from one city to another over great distances through the Earth. One interesting possibility would be to tunnel from New York to Leningrad, a distance of some 12,000 km over the Earth's surface or 108° of a great circle around the Earth.

Such a tunnel would reach a depth of 2640 km and would have to be impervious to the several thousand degree temperatures inside the Earth. However, if such a tunnel could be built, what would be the transit time from New York to Leningrad by a train traveling in an evacuated tunnel? Use the very interesting fact that gravity inside the Earth is nearly constant at 10 m/s² to a depth of 3000 km.

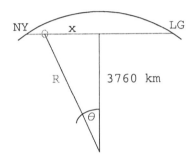

The train leaves the New York station at x= −5177 km

and arrives in Leningrad at x= +5177 km some time later.

At any time, the gravitational acceleration of the train along x is

$$\ddot{x} = -g\,\mathrm{Sin}\,\theta = -g\,\frac{x}{R} = -\frac{.010\,x}{\sqrt{x^2 + 3760^2}}$$

$$\ddot{x} = -\frac{.010\,x}{(x^2 + 3760^2)^{1/2}} \qquad x_0 = -5177\ \mathrm{km} \qquad \dot{x}_0 = 0$$

The travel time is found by using *Mathematica* to solve the above non-linear Differential Equation for x, and plotting the result versus time.

```
In[1]:=
    sol1 = NDSolve[{x"[t] == -.01*x[t] / (x[t]^2 + 3760^2)^0.5,
        x[0] == -5177, x'[0] == 0}, x, {t, 0, 6000}]
    Plot[Evaluate[-x[t] /. sol1, {t, 0, 1200}],
        AxesLabel → {"Time(s)", "Dist(km)"}];
    InterpFunc1 = x /. sol1[[1]]; InterpFunc2 = x' /. sol1[[1]];
    Plot[Evaluate[x'[t] /. sol1, {t, 0, 1200}],
        AxesLabel → {"Time(s)", "Vel(km/s)"}];
```

If we plot the distance travelled versus time, we find the train crosses the midpoint (x = 0) at t ≃ 1200 seconds.

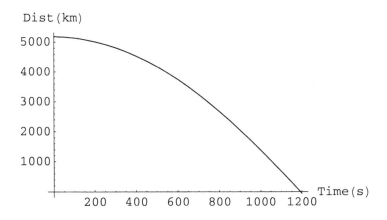

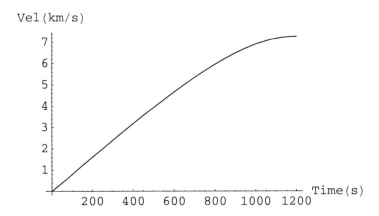

Therefore the travel time is twice 19.9 minutes, the time to reach x = 0, and is 39.8 minutes from NewYork to Leningrad. The velocity at center-point is over 7 kilometers/second, or about 16,000 miles per hour.

```
In[80]:= Table[ {t / 60, InterpFunc1[t], Chop[InterpFunc2[t]]},
             {t, 0, 2400, 120}] // TableForm
```

t (min)	x (km)	v (km/s)
0	-5177.	0
2	-5118.78	0.969671
4	-4944.59	1.93156
6	-4655.87	2.87685
8	-4255.25	3.7945
10	-3746.89	4.66942
12	-3137.15	5.47994
14	-2435.53	6.19429
16	-1656.16	6.76723
18	-819.341	7.14159
20	47.6908	7.26368
22	912.926	7.11212
24	1744.66	6.71313
26	2516.4	6.12208
28	3208.55	5.39526
30	3807.59	4.57636
32	4304.42	3.69585
34	4692.97	2.77454
36	4969.27	1.82694
38	5130.82	0.863801
40	5176.3	-0.106241

As a check, we may query the Interpolation Function values to find x and v at any time.

```
In[5]:= Print[ "At t=39.8 min x=", InterpFunc1[39.8 * 60], "km",
          "    At t=19.9 min, vmax=", InterpFunc2[19.9 * 60] /
          1.609 * 3600, " mph"]
```

```
At t=39.8 min x=5177.km     At t=19.9 min, vmax=16252.8 mph
```

The total travel time is 39.8 minutes from New York to Leningrad.

[World travelers will note that the name of the Russian city has changed from Leningrad to St. Petersburg.]

Problem 2-2 Lunar Subway

Assume the moon is of constant density. Find the maximum velocity and transit time from the North pole to the South pole of the moon. The radius of the moon R_{moon}= 1700 km. The surface gravity of the moon g_{moon} = 1.7 m/s^2.

The moon appears to be made of the same material as the crust of the Earth. Thus the assumption of a constant-density moon is not a bad one. Then the mass inside a distance x from the center of the moon is $m = \frac{4}{3}\pi\rho x^3$, and the gravitational acceleration of an object inside the moon is

$$\ddot{x} = -\frac{mG}{x^2} = -\frac{4}{3}\pi\rho xG = -g\frac{x}{R_{moon}}$$

and the equation we will solve with *Mathematica* is

$$\ddot{x} = -.0017\frac{x}{R_{moon}} \qquad x_o = -1700 \text{ km} \qquad \dot{x}_o = 0$$

```
In[1]:=  sol = DSolve[{x''[t] == -.0017/1700*x[t], x[0] == -1700, x'[0] == 0},
             {x[t], x'[t]}, t] // Chop
         Plot[Evaluate[-x[t] /. sol, {t, 0, 1560},
             AxesLabel → {"Time(s)", "Dist(km)"}]];
         x[t_] = x[t] /. sol[[1]]; Plot[Evaluate[x'[t] /. sol,
             {t, 0, 1560}], AxesLabel → {"Time(s)", "Vel(km/s)"}];
         v[t_] = x'[t] /. sol[[1]]; Print[" v(t)= ", v[t]]

Out[1]=  {{x[t] → -1700. Cos[0.001 t]}}
```

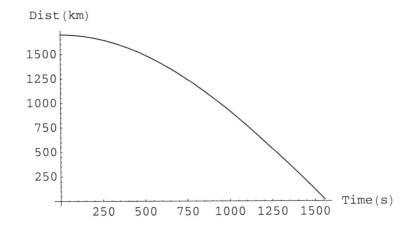

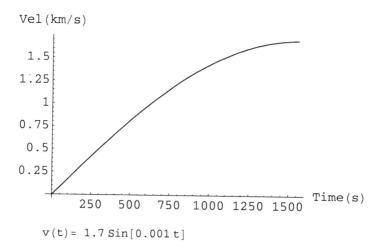

$$v(t) = 1.7 \, Sin[0.001 \, t]$$

If we Table the results, we find that the transit time from North-to-South pole of the moon is 52 minutes.

In[31]:= **Table[{t / 60, x[t], v[t]}, {t, 0, 3120, 120}] // TableForm**

t (min)	x (km)	v (km / s)
0	-1700.	0
2	-1687.77	0.203511
4	-1651.27	0.404094
6	-1591.02	0.598866
8	-1507.89	0.785025
10	-1403.07	0.959892
12	-1278.07	1.12095
14	-1134.69	1.26589
16	-974.984	1.39263
18	-801.258	1.49933
20	-616.008	1.58447
22	-421.898	1.64682
24	-221.72	1.68548
26	-18.3534	1.6999
28	185.277	1.68987
30	386.244	1.65554
32	581.654	1.5974
34	768.699	1.51628
36	944.689	1.41335
38	1107.09	1.2901
40	1253.57	1.14829
42	1382.02	0.989962
44	1490.59	0.817398
46	1577.72	0.633078
48	1642.16	0.439653
50	1682.99	0.239904
52	1699.6	0.0367047

The maximum velocity at $t = 26$ minutes is $v = 1.7$ km/s.

Problem 2-3 Speed Bump

Let's say we're cruising along the highway in our favorite automobile, when up ahead is a *speed bump*. Should we speed-up or slow down? You had best slow-down, if you wish to preserve the integrity of your car, and the following problem will show the reason.

Assume the highway engineers have placed a 0.8-meter long, 0.1 meter-high sinusoidal *speed bump* in your lane, and you approach the bump at 2 m/s, what is the suspension-system response?

Now take a computer-simulated run at the speed-bump at 4 m/s and at 8 m/s. What is the suspension-system response at each of these velocities? Assume m = 1000 kg, c = 3400 N-s/m, k = 80,000 N/m.

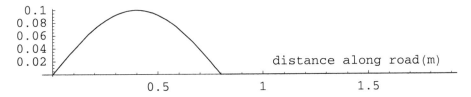

Here is how we see the speed-bump.

```
In[14]:=  L = 0.8; A = .10; v = 2; m = 1000; k = 80000; c = 3400;
          ω0 = √(k/m) ; ω = π * v / L;
          bump = Plot[.10 * Sin[π * x / L] * If[x ≤ 0.8, 1, 0],
            {x, 0, 2}, AspectRatio → 0.15,
            Prolog → {Text["distance along road(m)", {1.50, .022}]}]
          road = Plot[.10 * Sin[(π * v / L) * t] * If[t ≤ 0.4, 1, 0], {t, 0, 1},
            Prolog → {Text["time along road(s)", {.75, .028}]}]
```

```
0.1
0.08
0.06
0.04              time along road(s)
0.02

        0.2     0.4      0.6      0.8
```

-- Road surface vs time in seconds --

This is how the suspension system sees the road surface at a speed of 2 m/s.

Let's now look at the suspension system's response on encountering the speed-bump at v = 2 m/s.

We want to solve the system

$$m\,y'' + c\,y' + k\,y = k\,A\,\text{Sin}[\tfrac{\pi\,v}{L}\,t] + c\,\omega\,A\,\text{Cos}[\tfrac{\pi\,v}{L}\,t]\quad\text{where } t \le 0.4\ \text{s}$$

$$m\,y'' + c\,y' + k\,y = 0\quad\text{for } t \ge 0.4\ \text{s}$$

```
In[5]:=

L = 0.8; A = .10; v = 2; m = 1000; k = 80000; c = 3400; ω0 = √k/m ;
ω = π * v / L; sol1 = NDSolve[{m*y''[t] + c*y'[t] + k*y[t] ==
    k*A*Sin[(π*v/L)*t]*If[t ≤ 0.4, 1, 0] +
      c*ω*A*Cos[(π*v/L)*t]*If[t ≤ 0.4, 1, 0], y[0] == 0, y'[0] == 0},
  y, {t, 0, 1.0}]; InterpFunc1 = y /. sol1[[1]];
InterpFunc2 = y' /. sol1[[1]];
Acc[t_] = (k*A*Sin[ω*t]*If[t ≤ 0.4, 1, 0] + c*ω*A*Cos[ω*t]*
      If[t ≤ 0.4, 1, 0] - c*InterpFunc2[t] - k*InterpFunc1[t])/m;
plot2 = Plot[Evaluate[y[t] /. sol1], {t, 0, 1},
  PlotStyle → {{RGBColor[1, 0, 0]}}, AspectRatio → 0.2,
  AxesLabel → {"time(s)", "y(m)"}];
Show[{road, plot2}, PlotRange → All, AspectRatio → 0.2,
 AxesLabel → {"t(s)", "y(m)"}, Prolog →
  {{Text["Road", {0.05, 0.03}]}, {Text["Response", {.35, .12}]}}]
```

Road profile and response at velocity 2 m/s. Notice how the suspension system "carries" the front-end of the automobile up-and-over the bump, letting the front-end down gently as the rear wheels make contact with the "bump". Thus the "underbelly" of the auto will not make contact with the "bump".

```
In[9]:= Table[{t, Chop[InterpFunc1[t]], Chop[InterpFunc2[t]],
          Acc[t], Chop[.10 * Sin[(π * v / L) * t] * If[t ≤ 0.4, 1, 0]]},
          {t, 0, 0.9, .05}] // TableForm
```

t (s)	y (m)	Y' (m / s)	Acc	Road
0	0	0	2.67035	0
0.05	0.00429707	0.187594	4.54697	0.0382
0.1	0.0195637	0.423024	4.5417	0.0707
0.15	0.0457914	0.610445	2.67411	0.0923
0.2	0.0784193	0.667736	-0.54384	0.1
0.25	0.109583	0.547693	-4.25963	0.0923
0.3	0.130198	0.249815	-7.49657	0.0707
0.35	0.132356	-0.179307	-9.38444	0.0382
0.4	0.111457	-0.656432	-9.35502	0
0.45	0.0719014	-0.892601	-2.71727	0
0.5	0.0255245	-0.930394	1.12138	0
0.55	-0.018219	-0.793988	4.15708	0
0.6	-0.051828	-0.535292	5.96626	0
0.65	-0.070808	-0.22021	6.41342	0
0.7	-0.074013	0.0854375	5.63062	0
0.75	-0.063330	0.327851	3.95173	0
0.8	-0.042861	0.473066	1.82051	0
0.85	-0.017840	0.510009	-0.306808	0
0.9	0.00649635	0.448927	-2.04606	0

Let's now look at the suspension system response and the accelerations experienced in contacting the speed bump at 10 m/s.

```
Clear[x, t, L, A];
L = 0.8; A = .10; v2 = 10; m = 1000; k = 80000; c = 3400; ω = π * v2 / L; bump =
  Plot[.10 * Sin[π * x / L] * If[x ≤ 0.8, 1, 0], {x, 0, 2}, AspectRatio → 0.15,
    Prolog → {Text["distance along road(m)", {1.45, .03}]}]
road = Plot[.10 * Sin[(π * v2 / L) * t] * If[t ≤ 0.08, 1, 0], {t, 0, .2},
  AspectRatio → 0.15,
    Prolog → {Text["time along road(s)", {.145, .03}]}]
sol2 = NDSolve[{m * y''[t] + c * y'[t] + k * y[t] ⩵ k * A * Sin[(π * v2 / L) * t] *
        If[t < 0.08, 1, 0] + c * ω * A * Cos[(π * v2 / L) * t] * If[t < 0.08, 1, 0],
      y[0] ⩵ 0, y'[0] ⩵ 0}, y, {t, 0, 1.0}];
InterpFunc1 = y /. sol2[[1]]; InterpFunc2 = y' /. sol2[[1]];
InterpFunc3 = y" /. sol2[[1]];
plot3 = Plot[Evaluate[y[t] /. sol2],
    {t, 0, 0.8}, PlotStyle → {{RGBColor[1, 0, 0]}},
    AspectRatio → 0.2, AxesLabel → {"time(sec)", "y(m)"}];
Show[{road, plot3}, PlotRange → All,
  Prolog → {{Text["Road", {0.05, 0.03}]},
    {Text["Response", {.25, .06}]}, {Text["Time(s)", {0.6, .02}]}}]
```

Road profile and response at velocity 10 m/s. At the higher velocity, the suspension-system sees the speed bump as a speed "spike". Let's Table the data and see how much of a jolt the driver and passengers take on hitting the speed bump at 10 m/s.

```
In[43]:= Table[{t, InterpFunc1[t], Chop[InterpFunc2[t]], InterpFunc3[t],
          Chop[.10 * Sin[(π * v2 / L) * t] * If[t ≤ 0.08, 1, 0]]},
         {t, 0, .32, .02}] // TableForm
```

t (s)	y (m)	y' (m / s)	Acc (m / s²)	road
0	0	0	13.3518	0
0.02	0.002870	0.288809	13.8865	0.0707
0.04	0.010985	0.49432	5.44059	0.1
0.06	0.021136	0.478534	-7.1021	0.0707
0.08	0.028557	0.232136	-14.295	0
0.1	0.032576	0.169397	-3.1820	0
0.12	0.035326	0.105556	-3.1849	0
0.14	0.036804	0.042655	-3.0894	0
0.16	0.037051	-0.01742	-2.9048	0
0.18	0.036137	-0.07302	-2.6427	0
0.2	0.034169	-0.12270	-2.3163	0
0.22	0.031276	-0.16534	-1.9399	0
0.24	0.027608	-0.20007	-1.5284	0
0.26	0.023330	-0.22634	-1.0968	0
0.28	0.018613	-0.24390	-0.6597	0
0.3	0.013631	-0.25279	-0.2310	0
0.32	0.008557	-0.25329	0.17661	0

At the higher speed (10 m/s), the chassis doesnt have time to "clear" the speed bump... At this speed we are attempting to drive "right through" the speed bump. Not a good idea.

If there is any downward movement of the chassis, we will scrape the bottom of our auto. What will also happen at 10 m/s is that the driver will feel a "shock" of 1.4 gees as the front tires run up and over the "speed bump". Running over a speed bump at speeds greater than 20 mph would definitely jar your teeth and put terrific stress on your suspension system.

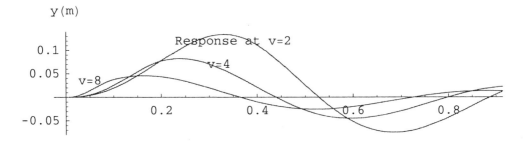

Because the suspension system takes a certain amount of time to respond, the suspension-system response to the speed bump is good at velocity 2 m/s, fair at 4 m/s, and terrible at 8 m/s.

Problem 2-4   RLC Circuit

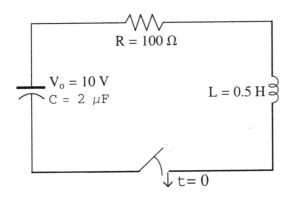

The differential equation for the RLC-series circuit is

$$LC \frac{d^2 V}{dt^2} + RC \frac{dV}{dt} + V = 0 \quad \text{where } V_o = 10 \text{ volts and } I_o = 0.$$

It is easy enough to plot the voltage response using **NDSolve**, however since this is a *linear equation*, we can utilize **DSolve** to obtain the complete equation of the voltage response versus time. Then we can just read the time-constant and frequency of oscillation of the circuit.

```
In[28]:=  Clear[v, t, R, L, c]; L = 0.5; R = 100; c = 2 * 10^-6;
          sol = DSolve[{.000001 v''[t] + .0002 v'[t] + v[t] == 0,
              v[0] == 10, v'[0] == 0}, v[t], t] // FullSimplify
          Plot[v[t] /. sol, {t, 0, .030}, PlotRange → {-8, 10}]
```

Out[29]= $\{\{v[t] \to e^{-100 \cdot t} (10. \cos[994.987 \, t] + 1.00504 \sin[994.987 \, t])\}\}$

The natural response of an RLC series circuit

Given that the voltage across the capacitor with time is given by

$$\{\{v[t] \rightarrow e^{-100.\,t}\ (10.\ Cos[994.987\ t] + 1.00504\ Sin[994.987\ t])\}\}$$

We find the time-constant from the exponential term: $e^{-t/\tau}$, $\tau = .01s = 10$ ms.
The oscillation frequency of this *underdamped* circuit is $f = \frac{\omega}{2\pi} = \frac{994.987}{2\pi}$
$= 158.4$ Hz.

Given the time-constant, we may plot the curve with an exponential envelope

```
Plot[{v[t] /. sol, 10 Exp[-100 t], -10 Exp[-100 t]}, {t, 0, .030},
  PlotRange → {-8, 10}, AxesLabel → {"T(s)", "Voltage"}]
```

Out[43]= $\{\{v[t] \rightarrow e^{-100.\,t}\ (10.\ Cos[994.987\ t] + 1.00504\ Sin[994.987\ t])\}\}$

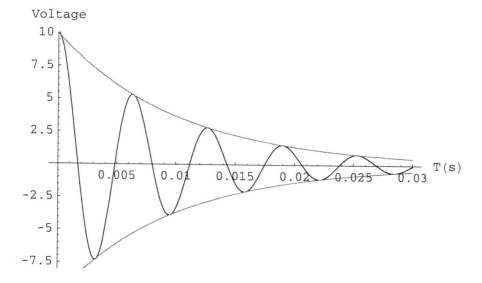

The response of the RLC-circuit with an exponential envelope.

As in Section 2-1, we may also determine the amount of damping per cycle.

Given that $V[t] = e^{-100.\,t}\ (10.\ Cos[994.987\ t] + 1.005\ Sin[994.987\ t])$

After one cycle, $994.987\ t = 2\pi$

Thus the time of one cycle is $t = .00631$ sec.

At that time, $e^{-100t} = e^{-.631} = .532$

Thus the voltage amplitude decreases by a factor of **Exp[-.631]** $= .532$
every cycle.

Problem 2-5 Impulse Response

The differential equation of a mechanical system is given by

$$y'' + 3\,y' + 2\,y = F(t) \qquad y_o = 0, \quad y_o' = 0.$$

This corresponds to a system with mass m = 1, spring constant k = 2, and damping coefficient c = 3, at rest at t = 0.

Let's find the system response to each FΔt

     (A) a unit impulse of (F=1 acting for Δt=1 second)
     (B) a unit impulse of (F=2 acting for Δt=0.5 second).
     (C) a unit impulse of (F=10 acting for Δt=0.1 sec.)
     (D) a unit impulse from the Dirac Delta function

(A) The unit impulse of FΔt=1, F=1 acting for 1 second

```
Clear[y];
q1 = Plot[{1 * UnitStep[t] - 1 * UnitStep[t - 1]},
    {t, -1.1, 2.75}, PlotRange → {0, 1.5}, AxesLabel → {None, "Force"}]
sol1 = NDSolve[{y''[t] + 3 y'[t] + 2 y[t] == UnitStep[t] - UnitStep[t - 1],
    y[0] == y'[0] == 0}, y, {t, -1.8, 4}];
r1 = Plot[y[t] /. sol1, {t, -1., 3.0}, PlotRange → {0, 0.28},
    AxesLabel → {"t(s)", "y"}]
```

In[24]:= **Show[GraphicsArray[{q1, r1}]]**

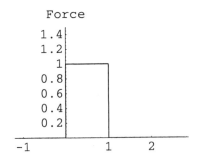

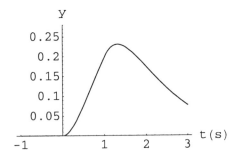

    The Impulse   **The Response**

(B) A force F=2 acting for 0.5 seconds

```
p2 = Plot[{2 * UnitStep[t] - 2 * UnitStep[t - .5]},
   {t, -1., 2.}, PlotRange → {0, 2.5}]
sol2 = NDSolve[
   {y''[t] + 3 y'[t] + 2 y[t] == 2 * UnitStep[t] - 2 * UnitStep[t - .5],
    y[0] == y'[0] == 0}, y, {t, -1.8, 4}];
r2 = Plot[y[t] /. sol2, {t, -1., 3.0}, PlotRange → {0, 0.28},
   AxesLabel → {"t(s)", "y"}]
```

In[57]:= **Show[GraphicsArray[{p2, r2}]]**

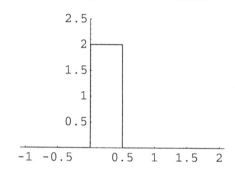

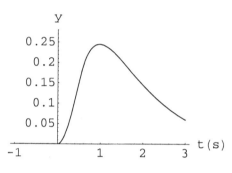

The Impulse   **The Response**

(C) The force is 10, the time is 0.1, the impulse is still one

```
Clear[y]; p10 = Plot[{10 * UnitStep[t] - 10 * UnitStep[t - .1]},
   {t, -1.0, 1}, PlotRange → {0, 10.2}]
sol10 = NDSolve[
   {y''[t] + 3 y'[t] + 2 y[t] == 10 * UnitStep[t] - 10 * UnitStep[t - .1],
    y[0] == y'[0] == 0}, y, {t, -1.8, 4}]
r10 = Plot[y[t] /. sol10, {t, -1., 3.0},
   PlotRange → {0, 0.28}, AxesLabel → {"t(s)", "y"}]
```

In[52]:= **Show[GraphicsArray[{p10, r10}]]**

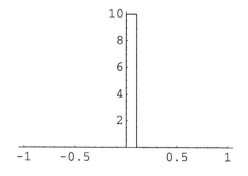

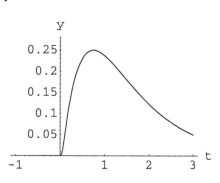

The Impulse   **The Response**

(D) The unit impulse from the Dirac Delta function: F→∞ as Δt → 0, but
FΔt =1. Note that with the Dirac function applied at t = 0, FΔt = mΔv =1 and
since m=1, the system must leave t =0 with a velocity y′ = 1. *We must modify
the initial conditions*

```
In[7]:= DSolve[{y''[t] + 3 y'[t] + 2 y[t] == DiracDelta[t],
           y[0] == 0, y'[0] == 1}, y[t], t]
         Plot[y[t] /. %, {t, -1.8, 3}, PlotRange → {0, .25}]
```

$$Out[8]= \{\{y[t] \to e^{-2t}\, (-1 + e^t)\, \text{UnitStep}[t]\}\}$$

Response of system to Dirac Delta function at t = 0. (Almost identical to Force of 10 for 0.1 s)

(D) The unit impulse from the Dirac Delta function F→∞ as Δt → 0, but
FΔt =1. Note that with the Dirac function applied at t =.000001s, we dont need
to worry about the initial conditions. The system is at rest at t =0, and the Dirac
function chimes in after that time. At t = 0, the system has position y = 0 and
velocity y′ = 0. *We dont need to modify the initial conditions.*

```
In[1]:= DSolve[{y''[t] + 3 y'[t] + 2 y[t] == DiracDelta[t - .000001],
           y[0] == 0, y'[0] == 0}, y[t], t]
         Plot[y[t] /. %, {t, -1.8, 3}, PlotRange → {0, .25}]
```

$$Out[2]= y[t] \to e^{-2t}\, (-1 + e^t)\, \text{UnitStep}[-1 \times 10^{-6} + t]$$

Response of system to Dirac-Delta function at t =10⁻⁶ s. (Indistinguishable from above curve)

Problem 3-1 Gravity Waves and Pulsar 1913+16

The equations of General Relativity may be used to find the power radiated by gravitational waves, the rate of decrease of the semi-major axis of the binary Pulsar, the rate of decrease of the orbital period, and the total time for system collapse.

$$G = 6.67 * 10^{-11}; \; c = 3 * 10^8; \; M1 = M2 = 1.4 * 2 * 10^{30};$$
$$M = M1 + M2; \; a = 1.95 * 10^9; \; P0 = 27900; \; \epsilon = .617;$$

(* Note: for $\epsilon = .617$, we need η *)

$$\eta = 1 / (1 - \epsilon^2)^{7/2} * \left(1 + \frac{73}{24} \epsilon^2 + \frac{37}{96} \epsilon^4\right)$$

11.8428

A. Power emitted in gravitational waves (in Watts)

$$P = 6.4 * (G^4 * (M1 ^ 2 * M2 ^ 2) * (M1 + M2)) / (a^5 * c^5) * 11.84$$

7.53487×10^{24}

B. Rate of decrease of semi-major axis of system (in meters per year)

$$adot = \frac{64 * G^3 * (M1 * M2) * M}{5 * a^3 * c^5} * 11.84 * 3.156 * 10^7$$

3.45836

C. Rate of decrease of period of binary pulsar system (seconds per year)

$$Pdot = \frac{96 * G^3 * (M1 * M2) * M}{5 * c^5} * \left(\frac{4 \pi^2}{G * M}\right)^{4/3} * \frac{11.84}{(27900)^{5/3}} * 3.156 * 10^7$$

0.000074895

D. Time for system to in-spiral and collapse (in years)

$$T = \frac{12 * (1.55 * 10^9)^4}{3.156 * 10^7 * 19 * 6.86 * 10^{19}} \int_0^{.617} x^{29/19} \frac{(1 + (121. / 304) x^2)^{.5137}}{(1 - x^2)^{1.5}} \, dx$$

3.04533×10^8

Note that the results from General Relativity theory account for 98 per cent of the observed value of 76.5 μs orbital period decrease per year.

For more details on gravitational radiation, see
Charles Misner, Kip Thorne, and John A. Wheeler (1973) *Gravitation* [Freeman]
Bradley Carroll, Dale Ostlie (1996) *Introduction to Modern Astrophysics* [Add – Wesley]

Certain constants β, η, k for the PSR 1913+16 system are evaluated as follows from the 1964 paper by Philip C. Peters, "Gravitational Radiation from Two Point Masses" *Physical Review* <u>136</u>, B1224-1232.

$$G = 6.67 * 10^{-11}; \quad c = 3 * 10^{8}; \quad M1 = M2 = 2.8 * 10^{30}; \quad M = M1 + M2;$$

$$\beta = \frac{64}{5} \frac{G^3 * M1 * M2 * (M1 + M2)}{c^5}$$

6.86255×10^{19}

$In[76]:=$ $\epsilon = .617;$

$$\eta = \frac{1}{(1 - \epsilon^2)^{7/2}} * \left(1 + \frac{73}{24} \epsilon^2 + \frac{37}{96} \epsilon^4\right)$$

$Out[77]=$ 11.8428

For this particular binary pulsar system, it is necessary to find the constant k in the equation

$$a(\epsilon) = k * \frac{\epsilon^{12/19}}{(1-\epsilon^2)} * \left(1 + \frac{121}{304} \epsilon^2\right)^{870/2299}$$

At the present time, $a = 1.95 \times 10^9$ meters and $\epsilon = .617$, and to evaluate k, we have

$$NSolve\left[1.95 * 10^9 == k * \frac{\epsilon^{12/19}}{(1 - \epsilon^2)} * \left(1 + \frac{121}{304} \epsilon^2\right)^{870/2299}, k\right]$$

$\{\{k \to 1.55248 \times 10^9\}\}$

Problem 3-2 Halley's Comet Equations

Use the equations of Section 3.3 to find all the orbital parameters of Halley's comet, given only the observed velocity of 54.45 km/s at closest approach to the Sun, which was 0.59 AU on its last approach in 1986. The mass of the Sun is 2×10^{30} kg.

Let us use the following equations from Section 3.3

$\quad\quad$ (1) perihelion velocity $\quad\quad\quad v_p = \frac{2\pi a}{T} \sqrt{\frac{1+\epsilon}{1-\epsilon}}$

$\quad\quad$ (2) the Newton-Kepler Law $\quad T^2 = \frac{4\pi^2 a^3}{G(M1+M2)}$

$\quad\quad$ (3) perihelion distance $\quad\quad\quad d_p = a(1-\epsilon)$

Notice first, *and this is very important,* that the first of these equations contains *two* singular points -- at $\epsilon = 1$, and $T = 0$. Now you and I would never choose these values for eccentricity and time period, but the computer is looking at all possible values for a solution. It is best, therefore, to put Equations (2) and (3) into Equation (1).

Thus, $\quad v_p = \sqrt{GM} \sqrt{\frac{1+\epsilon}{d_p}}$ and we now use *Mathematica* to solve for ϵ with *perihelion* velocity $v_P = 54.45$ km/s and distance $d_P = 88.5 \times 10^6$ km.

```
In[1]:= Clear[ε, a, T, d];
        vp = 54.45; d = 88.5 * 10^6; G = 6.67 * 10^-20; M = 2 * 10^30;
        (* note dist in km, time in seconds*)
        sol = NSolve[vp == √GM * √(1+ε) / d, ε]

Out[4]= {{ε → 0.966904}}
```

We now solve for the orbital time T and the semi-major axis a

```
In[22]:= ε = .967;
         NSolve[{d == a * (1-ε), T == √4π² * a^3 / (G*M) }, {a, T}]

Out[23]= {{a → 2.68182 × 10^9, T → 2.38916 × 10^9}}
```

In terms of AU and years,

```
In[28]:= Print["Au= ", 2.68182 * 10^9 / (1.5 * 10^8),
         " Tyears= ", 2.38916 * 10^9 / (3.156 * 10^7)]

         Au= 17.8788  Tyears=75.7022
```

Since we now have the main parameters ϵ, T, and a, it is relatively easy to find the comet's distance and velocity at Aphelion

```
In[31]:= a = 17.88; ε = .967;

       NSolve[{da == a * (1 + ε), va == vp * (1 - ε)/(1 + ε)}, {da, va}]

Out[32]= {{da → 35.17, va → 0.913498}}
```

Table of values for Halley's comet

Semi-major axis **a**	17.88 AU	Orbital Time **T**	75.7 years
Perihelion distance	0.59 AU	Perihelion velocity	54.45 km/s
Aphelion distance	35.17 AU	Aphelion velocity	0.91 km/s
Orbital eccentricity	$\epsilon = .967$		

```
Needs["Graphics`Graphics`"];
a = 17.88; ε = .967;
comet = PolarPlot[a * (1 - ε²) / (1 - ε * Cos[θ]), {θ, 0, 2 π}]
```

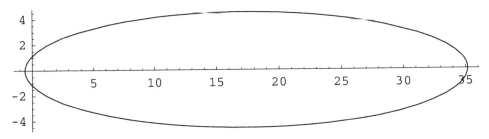

The orbit of Halley's comet. $r = \dfrac{a\,(1-\epsilon^2)}{1 + \epsilon \, \cos \theta}$ a = 17.88 AU, $\epsilon = .967$

In 1986, the spacecraft *Giotto* approached to within 600 km of the nucleus of Halley's comet. From the deviation in its trajectory, the mass of comet Halley was estimated at about 10^{15} kg. Halley's comet has about the same mass as Mt Everest.

Problem 3-3 The Largest Stars

Plaskett's binary star system consists of two stars that revolve in a circular orbit about a center of gravity midway between them. This means the masses of the two stars are equal.

If the orbital velocity of each star is 220 km/s and the orbital period of each is 14.4 days, find the mass M of each star.

For a circular orbit, $\quad v = \frac{2 \pi r}{T}$

the radius of the orbit $\quad r = \frac{v\,T}{2\,\pi} = \frac{(220)\ (14.4)\ (86400)}{2\,\pi} = 43.56 \times 10^6$ km

The interstellar separation $\quad a = r + r = 87.12 \times 10^6$ km

```
r = 43.56 * 10^6; Needs["Graphics`Graphics`"];
stars = PolarPlot[r, {t, 0, 2 π}, PlotStyle → Dashing[{.03}],
   AspectRatio → 1, AxesStyle → GrayLevel[0.7],
   PlotRange → {{-5 * 10^7, 5 * 10^7}, {-5 * 10^7, 5 * 10^7}},
   Prolog → {{Hue[.67], Circle[{-4.35 * 10^7, 10}, 6000000]},
   {Hue[0.67], Circle[{4.35 * 10^7, 100}, 6000000]}}]
```

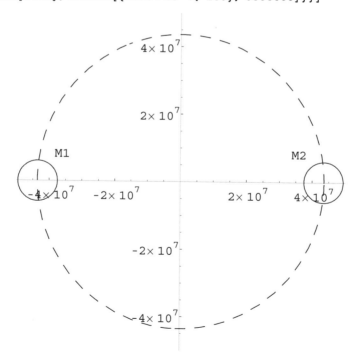

The orbits of the two stars comprising Plaskett's binary system.
The radius r = 43.56 × 10⁶ km, and the separation of the two stars is
a = 87.12 × 10⁶ km.

We may now use the Newton-Kepler Law to find the masses of the stars.

$$T^2 = \frac{4\,\pi^2\,a^3}{G\,(M1 + M2)}$$

```
In[50]:=  (* dist in km, time in s, mass in kg *)
          a = 87.12 * 10^6;  T = 14.4 * 86400;  G = 6.67 * 10^-20;
          sol = Solve[T^2 == 4 * π^2 * a^3 / (G * 2 M), M]

Out[51]=  {{M → 1.26417 × 10^32}}


In[57]:=  Print["The mass of each star, in solar masses is  M →",
          (M / (2 * 10^30)) /. sol ]

          The mass of each star, in solar masses is  M →{63.2084}
```

Therefore the total mass of Plaskett's binary system is approx 126 solar masses. The two blue-giant stars comprising Plaskett's system are among the most massive known.

Problem 3-4 The Alaska Pipeline

For the above-ground portion of the Alaska pipeline, the 1.2 meter-diameter pipe is wrapped with 10-cm of fiberglass insulation of thermal conductivity k = 0.035 W/m·°C.

We will find the heat loss per meter of pipeline when the oil inside the pipe is at 50°C, and the air temperature outside is – 40 °C with a convective heat transfer coefficient of h = 12 W/ m²·°C.

Solution to Alaska Pipeline problem:

The rate at which heat flows through the insulation is equal to the rate at which heat is abstracted away by the air currents at the surface of the pipeline. Therefore

$$\dot{Q} = \frac{2\,\pi\,k\,L\,(T_i - T_2)}{\ln\,(r_o/r_i)} = 2\pi r_o h\,L\,(T_2 - T_{air})$$

We are solving for T_2 the temperature at the surface of the pipeline

```
In[3]:=  k = .035; Ri = 0.60; Ro = 0.70; L = 1; h = 12; Ti = 50; TA = -40;
         sol = Solve[
           k * 2 Pi * 1 * (Ti - T2) / Log[Ro / Ri] == h * 2 Pi * Ro * 1 * (T2 - TA), T2]

Out[4]=  {{T2 → -37.6313}}
```

The temperature at the surface of the pipeline (with insulation) is just 2.4 degrees above the outside air temperature ! With or without gloves on, we wouldn't be able to detect the difference.

Now for the amount of heat lost per meter of pipeline

$$\dot{Q} = 2\pi r_o h\,L\,(T_2 - T_{air}) = 2\pi\,(0.7\text{ m})(12\text{ W/ m}^2\text{·°C})(1\text{ m})(2.4\text{ °C}) = 126\text{ W}$$

```
In[14]:=  Qdot = 2 π * .7 * 12 * 1 * (-37.6 - TA)

Out[14]=  126.669
```

This is not really significant. If we take the specific heat of the oil as C_V =2000 J/kg·°C then the average temperature drop ΔT in the oil (which is moving at 3 m/s) in one meter due to heat loss is

$$\rho\,A L\,C_V\,\Delta T = (126\text{ W})\,(1/3\text{ s}) = 42\text{ J}\quad(\rho = 870\text{ kg/m}^3,\ A = \pi\,r_i^2,\ L = 1\text{ m})$$

$$\Delta T = \frac{42\ J}{\rho\ AL\ C_V} = 21 \times 10^{-6}\ °C/m\ \text{ or }\ 2.1°C\ \text{ in }\ 100\ \text{km !!}$$

In[15]:= $\Delta T = 42 / (870 * \pi * .6^2 * 2000)$

Out[15]= 0.0000213426

And this is on the coldest of cold nights. It is interesting to see what would happen if the pipe-line wasn't insulated. Then the heat loss would be due to <u>both</u> convection and radiation. And for 0.5" stainless steel with k = 15 W/m·°C and emissivity 0.4, we can solve the following very difficult equation

$$\dot{Q} = \frac{2\ \pi\ k\ L\ (T_i - T_2)}{\ln\ (r_o/r_i)} = 2\pi r_o h\ L\ (T_2 - T_{air}) + \epsilon\ \sigma\ 2\pi r_o L\ (T_2^4 - T_{air}^4)$$

Using *Mathematica* we find $T_2 = 321.994$ K (or 48.99 °C). (Good thing there's insulation!)

In[13]:=   k2 = 15; σ = 5.67 * 10^-8; Ti = 323; TA = 233; Ri = 0.6; Ro = 0.612;
          sol = Solve[2 π * k2 * 1 * (Ti - T2) / Log[Ro / Ri] ==
          h * 2 π * Ro * 1 * (T2 - TA) + .4 * σ * 2 π * Ro * 1 * (T2 ^ 4 - TA ^ 4), T2]

Out[14]= {{T2 → -3907.14}, {T2 → 321.994},
          {T2 → 1792.57 - 3301.22 i}, {T2 → 1792.57 + 3301.22 i}}

Then, $\dot{Q}/L = 2\pi r_o h\ (321.994 - 233) + \epsilon\ \sigma\ 2\pi r_o\ (321.994^4 - 233^4)$

$\dot{Q}/L = 4107\ W\ +\ 680\ W\ =\ 4787$ Watts per meter

The heat loss per meter due to both convection and radiation for uninsulated pipe is

In[18]:= **Qdot = h * 2 π * Ro * 1 * (321.994 - 233) +**
          .4 * σ * 2 π * Ro * 1 * (321.994^4 - 233^4)

Out[18]= 4786.96

Problem 3-5 Counterflow Heat Exchanger

In a *counterflow* heat exchanger, the cold fluid enters from the left and exits heated on the right, and the hot fluid enters from the right and exits after cooling on the left. If hot oil at 200 °C with a mass flow rate of 3 kg/s and a specific heat of 1900 J/kg ·°C enters the outer shell from the right, and cool water at 20 °C with a mass flow rate of 0.9 kg/s and specific heat 4180 J/kg ·°C enters the inner tube from the left, what is the rate of heat exchange in this heat exchanger? We will keep the same heat-exchange parameters as in Section 3-5, the heat-transfer coefficient U = 460 W/m².°C, and the heat-exchange surface area is A = π D L. Where D = 10 cm, and the length is L = 40 meters.

Be sure to compare the effectiveness of this counter-flow heat exchanger with the effectiveness of the parallel-flow heat exchanger of Section 3-5. Which system does the better job of extracting heat from the hot fluid ?

Building the Differential Equations

For a heat exchanger, the cold fluid is heated dT by absorbing heat energy dQ through an area π D dx.

In the inner tube,

$$dQ = (\dot{m}\, c)_c\, dT = U\, \pi\, D\, (T^* - T)\, dx \qquad \text{Cold side}$$

Heat Absorbed = Heat Transferred

In the outer shell, the hot fluid (which is moving along – dx) *loses* heat energy dQ through the same area.

In the outer shell,

$$(\dot{m}\, c)_h\, dT^* = -\, U\, \pi\, D\, (T^* - T)\, (-\, dx) \qquad \text{Hot side}$$

Heat Lost = Heat Transferred

When we write the differential equations, with T for the cold fluid and T* for the hot fluid,

$$(\dot{m}\, c)_c\, \frac{dT}{dx} = U\, \pi\, D\, (T^* - T) \qquad \text{and}$$

$$(\dot{m}\, c)_h\, \frac{dT^*}{dx} = U\, \pi\, D\, (T^* - T)$$

The *Mathematica* program for the counterflow heat exchanger is

```
(* T[x]= water Temp,  S[x] = oil temp *)
mh = 3.0*1900; mc = 0.9*4180; A = 12.56; D0 = .10; L = 40; U = 460;
sol = NDSolve[{mc*T'[x] == U*π*D0*(S[x] - T[x]),
    mh*S'[x] == U*π*D0*(S[x] - T[x]),
    T[0] == 20, S[40] == 200}, {S, T}, {x, 0, 40.0}]
InterpFunc1 = S /. sol[[1]]; InterpFunc2 = T /. sol[[1]];
Table[{x, Chop[InterpFunc1[x]], Chop[InterpFunc2[x]]},
    {x, 0, 40, 5}] // TableForm
Plot[{S[x] /. sol, T[x] /. sol}, {x, 0, 40}, Prolog →
    {Text["Cold Fluid Out 140°C", {31, 124}], Text["→", {15, 60}],
     Text["←", {25, 165}], Text["Hot Fluid In 200°C", {32, 184}],
     Text["Cold Fluid In 20°C", {8.5, 34}],
     Text["Hot Fluid Out 120°C", {8.5, 110}]},
    Frame → True, GridLines → Automatic];
```

The *Mathematica* program for the Parallel-flow heat exchanger is

```
(* T[x]= water Temp,  S[x] = oil temp *)
mh = 3.*1900; mc = .9*4180; A = 12.56; D0 = .10; L = 40; U = 460;
sol = NDSolve[{mc*T'[x] == U*π*D0*(S[x] - T[x]),
    mh*S'[x] == - U*π*D0*(S[x] - T[x]),
    T[0] == 20, S[0] == 200}, {S, T}, {x, 0, 80.0}]
InterpFunc1 = S /. sol[[1]]; InterpFunc2 = T /. sol[[1]];
Table[{x, Chop[InterpFunc1[x]], Chop[InterpFunc2[x]]},
    {x, 0, 40, 5}] // TableForm
Plot[{S[x] /. sol, T[x] /. sol}, {x, 0, 40},
    Prolog → {Text["Oil", {15, 164}], Text["Water", {23, 90}]},
    AxesLabel → {"Dist(m)", "Temp(°C)"},
    Frame → True, GridLines → Automatic];
```

Note that the only difference between the two programs is the direction of flow of the oil. We now examine the performance of the two heat exchangers.

The Counterflow Heat Exchanger

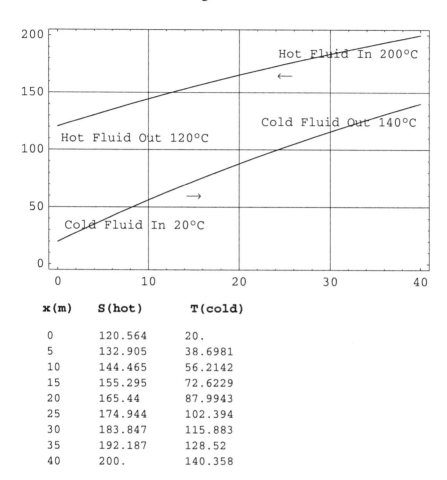

x(m)	S(hot)	T(cold)
0	120.564	20.
5	132.905	38.6981
10	144.465	56.2142
15	155.295	72.6229
20	165.44	87.9943
25	174.944	102.394
30	183.847	115.883
35	192.187	128.52
40	200.	140.358

Hot-side temperature declines from 200 °C to 120 °C while cold-side temperature increases from 20 °C to 140 °C. The counterflow set-up for the heat exchanger allows the cold fluid (water) to heat up to a temperature *greater than the temperature of the outgoing hot fluid* (oil). Let us see how much heat is extracted from the system in the counterflow mode:

$$Q = (\dot{m}\, c)_c\ \Delta T_c\ =\ (3762)\, (120)\ \text{W}\ =\ 451.4\ \text{kW}$$

The *effectiveness* ϵ in removing heat from the system is

$$\epsilon\ =\ \frac{\Delta T_c}{(T_h - T_c)_{max}}\ =\ \frac{120}{180} = 66.7\ \%$$

Let us compare this commendable performance with the parallel-flow system of Section 3.5. We keep everything the same, except the direction of flow, so that the hot fluid and the cold fluid both enter on the left-hand-side and both exit on the right-hand-side.

The Parallel-flow Heat Exchanger

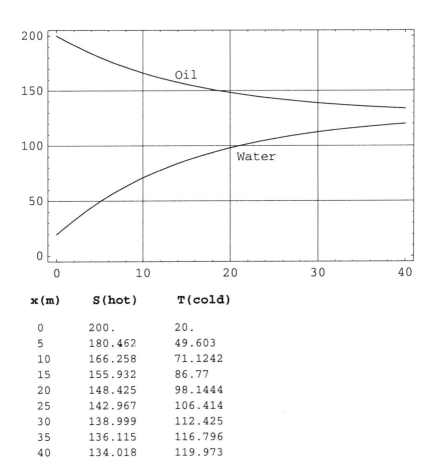

x(m)	S(hot)	T(cold)
0	200.	20.
5	180.462	49.603
10	166.258	71.1242
15	155.932	86.77
20	148.425	98.1444
25	142.967	106.414
30	138.999	112.425
35	136.115	116.796
40	134.018	119.973

Hot-side temperature declines from 200 °C to 134 °C while cold-side temperature increases from 20 °C to 120 °C. The parallel-flow set-up for the heat exchanger allows the cold fluid (water) to heat up to a temperature *less than* the temperature of the outgoing hot fluid (oil). Let us see how much heat is extracted from the system in the parallel-flow mode:

$$Q = (\dot{m}\,c)_c \;\Delta T_c \;=\; (3762)\,(100)\ \text{W} = 376.2\ \text{kW}$$

The *effectiveness* ϵ in removing heat from the system is

$$\epsilon \;=\; \frac{\Delta T_c}{(T_h - T_c)_{max}} \;=\; \frac{100}{180} = 55.6\ \%$$

The counterflow arrangement for a heat exchanger is more effective in removing heat from the system by approximately 11 %.

An interesting visual comparison between counterflow and parallel-flow heat exchangers is that if we reverse the direction of flow of the oil, then we change the performance by 11 percent. The counterflow heat exchanger does better in removing heat from the oil, because of its more uniform temperature gradient.

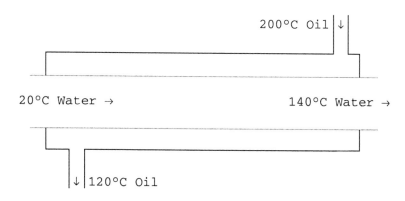

A counterflow heat exchanger. The water is heated from 20 °C to 140 °C, while the oil is cooled from 200 °C to 120 °C.

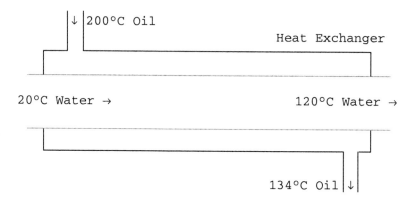

A parallel-flow heat exchanger. The water is heated from 20 °C to 120 °C, while the oil is cooled from 200 °C to 134 °C.

Problem 3-6 Women's Track and Field

Following, is a list of world records for women in running, from 100-meters to 10,000 meters.

Distance (m)	Time (s)	Avg Velocity (m/s)	Record Holder
100	10.49	9.53	Florence Joyner (USA) 1988
200	21.34	9.37	Florence Joyner (USA) 1988
400	47.60	8.40	Maria Koch (GER) 1985
800	113.28	7.06	Jarmila Kratochvilova (CZ) 1983
1000	149.34	6.69	Maria Mutola (MOZ) 1995
1500	230.46	6.51	Qu Yunxia (CHINA) 1993
1 mile	252.56	6.37	Svetlana Mastercova (RUS) 1996
2000	325.36	6.15	Sonia O'Sullivan (IRE) 1994
3000	486.11	6.17	Wang Junxia (CHINA) 1993
5000	864.53	5.78	Meseret Defar (ETH) 2006
10000	1771.78	5.64	Wang Junxia (CHINA) 1993

Let us first of all put this data into a form where we can see the average velocity to run a race of a given distance. *Mathematica* can manipulate and plot data in an easy to read form.

```
In[157]:= velocity = {{100, 9.53}, {200, 9.37}, {400, 8.40}, {800, 7.06},
            {1000, 6.70}, {1500, 6.51}, {1609, 6.37}, {2000, 6.15},
            {3000, 6.17}, {5000, 5.78}, {10000, 5.64}, {20000, 5.09}};
        velc = ListPlot[velocity, PlotRange → {{-100, 5100}, {5, 10}}]
```

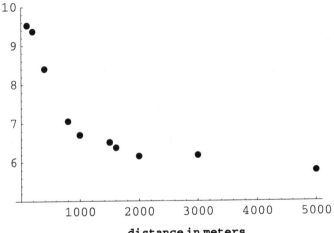

Let us apply the Joseph Keller theory to see how well we can match the track records for women. We choose the following physiological constants as representative for a female world-class runner

$F = 1.20 * 9.8$ m/s^2 (maximum acceleration of runner is 1.2 g)

$k = 1.14$ s^{-1} (internal and external resistance increases with velocity)

$\sigma = 36.5$ W/kg (rate of oxygen resupply to muscles)

$E_o = 2300$ J/kg (initial energy and oxygen supply in muscles)

We will apply these constants to the differential equations we solved before

$$\frac{dv}{dt} = f - k\,v \quad \text{and} \quad \frac{dE}{dt} = \sigma - f\,v$$

with solutions

for x < 300 m, $\quad v1 = \frac{F}{k}(1 - e^{-kt})$ and $x1 = \frac{F}{k}\,t - \frac{F}{k^2}(1 - e^{-kt})$ all acceleration

for x > 300 m, $\quad f = k\,v$ and $\frac{dE}{dt} = \sigma - k\,v^2$ constant velocity

$$z = x1 + v \cdot t2 \quad \text{and} \quad E2 = (k\,v^2 - \sigma) \cdot t2$$

```
In[185]:=
    Clear[x]; F = 1.2*9.8; k = 1.14; s = 36.5; E0 = 2300;
    sol1 = DSolve[{x"[t] == F - k*x'[t], x[0] == x'[0] == 0}, x[t], t] //
        FullSimplify
    sol2 = DSolve[{v'[t] == F - k*v[t], v[0] == 0}, v[t], t] //
        FullSimplify
Out[186]=  {{x[t] → -9.04894 + 9.04894 e^{-1.14 t} + 10.3158 t}}

           {{v[t] → 10.3158 - 10.3158 e^{-1.14 t}}}
```

Again, t is the acceleration time, and $t2$ is the time at constant velocity. E1 is the energy expended in the acceleration phase, and E2 is the energy expended in the run-phase.

We will solve for the time $t2$ at constant velocity, subject to the condition that all energy is used-up by the end of the race. Then, the total distance travelled is $z = x1 + x2 = x1 + v1 \cdot t2$ and the total time is $t + t2$. The average velocity for running the race is then $w = \frac{z}{(t+t2)}$.

Also, for races less than 300 m, there is only the acceleration phase, and the distance travelled is $x1$ in time t.

```
In[188]:=  F = 1.2 * 9.8; k = 1.14; s = 36.5; E0 = 2300;
           x1[t_] = (F / k) * t - (F / k^2) * (1 - Exp[-k * t])
           v1[t_] = (F / k) * (1 - Exp[-k * t])
           E1[t_] = F * x1[t] - s * t ;
           E2[t_] = E0 - E1[t] ;
           (* and since   E2 = (kv^2 - s) t2   *)
                          E0 - E1[t]
           t2[t_] = ─────────────── ;
                      (k * v1[t]^2 - s)
           z[t_] = v1[t] * t2[t] + x1[t] ;
           t3[t_] = t2[t] + t ;
           w[t_] = z[t] / t3[t] ;
           (*Table[{z[t],v1[t],t,t3[t]},{t,.737,1.637,.03}]//
             TableForm*)
           p = ParametricPlot[{z[t], w[t]}, {t, 0.735, 4.2},
               PlotRange → {{0, 5500}, {5, 10}}];
           r = ParametricPlot[{x1[t], x1[t] / t}, {t, 3, 26.8},
               PlotRange → {{0, 5500}, {5, 10}}];
```

The distance run and the velocity gained in the acceleration time t are

$Out[188]=$ $x[t] \rightarrow -9.04894 (1 - e^{-1.14 t}) + 10.3158 t$

$v[t] \rightarrow 10.3158 (1 - e^{-1.14 t})$

Let's find the total time $(t + t2)$ required to run each race using the Keller theory

```
(*Predicted times at various race distances -- *)
sol100 = FindRoot[100 == x1[t], {t, 10}];
sol200 = FindRoot[200 == x1[t], {t, 20}];
sol400 = FindRoot[400 == v1[t] * t2[t] + x1[t], {t, 1}];
t400 = (t2[t] + t) /. sol400;
sol800 = FindRoot[800 == v1[t] * t2[t] + x1[t], {t, 1}];
t800 = (t2[t] + t) /. sol800;
sol1000 = FindRoot[1000 == v1[t] * t2[t] + x1[t], {t, 1}];
t1000 = (t2[t] + t) /. sol1000;
sol1500 = FindRoot[1500 == v1[t] * t2[t] + x1[t], {t, 1}];
t1500 = (t2[t] + t) /. sol1500;
sol1609 = FindRoot[1609 == v1[t] * t2[t] + x1[t], {t, 1}];
t1609 = (t2[t] + t) /. sol1609;
sol2000 = FindRoot[2000 == v1[t] * t2[t] + x1[t], {t, 1}];
t2000 = (t2[t] + t) /. sol2000;
sol3000 = FindRoot[3000 == v1[t] * t2[t] + x1[t], {t, 1}];
t3000 = (t2[t] + t) /. sol3000;
sol5000 = FindRoot[5000 == v1[t] * t2[t] + x1[t], {t, 1}];
t5000 = (t2[t] + t) /. sol5000;
sol10000 = FindRoot[10000 == v1[t] * t2[t] + x1[t], {t, 1}];
t10000 = (t2[t] + t) /. sol10000;
Print[" t100 = ", t /. sol100];
Print[" t200 = ", t /. sol200]; Print[" t400 = ", t400];
Print[" t800 = ", t800]; Print[" t1000 = ", t1000];
Print[" t1500 = ", t1500]; Print[" t1609 = ", t1609];
Print[" t2000 = ", t2000]; Print[" t3000 = ", t3000];
Print[" t5000 = ", t5000]; Print["t10000 = ", t10000]
```

```
                t100  = 10.5711

                t200  = 20.2649

                t400  = 46.431

                t800  = 113.762

               t1000  = 148.404

               t1500  = 235.819

               t1609  = 254.953

               t2000  = 323.705

               t3000  = 499.953

               t5000  = 853.025

              t10000  = 1736.38
```

```
In[207]:= Show[{velc, p, r}, AxesLabel → {"Dist(m)", "Velocity(m/s)"}]
```

Let's run the *Mathematica* program and see how well we do

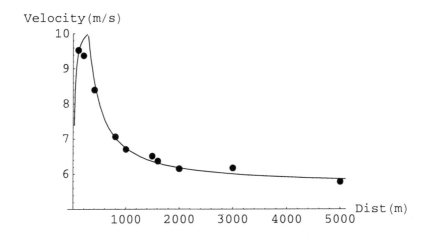

This appears to be a rather good fit, with only two dots away from the theory line (at 200 meters and 3000 meters). Assuming the theory to hold, we may expect the 200-m record to be broken, and the 3000-m record to stand for a number of years. The last question is how well do the theoretical and actual record numbers compare?

Track and Field World Records for Women 2007

Dist (m)	Theory Time	Record Time (s)	% Difference	
100	10.57	10.49	− 0.7	
200	20.26	21.34	4.8	(world record most likely to fall)
400	46.4	47.60	2.5	
800	113.7	113.28	− 0.4	
1000	148.4	149.34	0.6	
1500	235.8	230.46	− 2.3	
1 mile	254.9	252.56	− 0.6	
2000	323.7	325.36	0.9	
3000	499.9	486.11	− 2.8	(world record least likely to fall)
5000	853.0	864.53	1.3	
10000	1736.4	1771.78	2.0	

All the track records for women may be accounted for with the Physics model of running up to 10,000 meters. Beyond this range, when running fatigue sets in -- as the runner's system tries to clear the lactic acid build-up, acidosis of the muscles begins to occur -- this is often referred to as THE WALL. Many long-distance runners have learned to cope with the fatigue by training and diet, and also by running at a somewhat reduced pace. This can be accounted for in the Physics model by allowing k (the internal resistance) to increase for runs beyond 10,000 meters.

Problem 3-7 Black Hole Sun

At the center of the Milky Way, at least a dozen stars orbit an unseen object. One of these stars, named S2, completes an orbit in 15 years and approaches within 17 lighthours of the center. The spectrum of S2 is that of a main-sequence star 15 times the mass of the sun.

If the maximum excursion of S2 from the center is 10 lightdays, then what is the mass of the unseen object ?

We may use the Newton-Kepler Law to find the mass of the unseen object

$$T^2 = \frac{4\,\pi^2\,a^3}{G\,(M1 + M2)}$$

```
In[25]:=  (* dist in m, time in s, mass in kg *)
          c = 3 * 10^8;  T = 15 * 3.15 * 10^7;  G = 6.67 * 10^-11;
              c * (17 * 3600 + 10 * 86400)
          a = ──────────────────────────────── ;  M1 = 15 * 2 * 10^30;
                           2
          sol = Solve[T^2 == 4 * π^2 * a^3 / (G * (M1 + M2)), M2]

Out[27]=  {{M2 → 7.08613 × 10^36}}
```

```
In[41]:=    7.08 * 10^36
          ─────────────
            2 * 10^30

Out[41]=  3.54 × 10^6
```

The mass of the unseen object (a black hole) is about 3.5 million solar masses.

It is interesting to plot the orbit of star S2 about the galactic center

```
In[45]:= d = c * 17 * 3600.;
         sol2 = Solve[d == a * (1 - ε), ε]
```

```
Out[46]= {{ε → 0.867704}}
```

```
In[53]:= Needs["Graphics`Graphics`"];
              c * (17 * 3600 + 10 * 86400)
         a = ─────────────────────────── ; ε = .8677;
                   2 * (1.5 * 10¹¹)
         S2 = PolarPlot[a * (1 - ε²) / (1 - ε * Cos[θ]), {θ, 0, 2 π}]
```

Orbit of star S2 about the galactic center. All distances in AU.

One final thought, how close does S2 come to the event horizon (the Schwarzschild radius) ?

```
In[64]:= (* dist in m, time in s, mass in kg *)  M = 3.5 * 10⁶ * 2 * 10³⁰;
              2 G * M
         r = ───────  ╱ (1.5 * 10¹¹)
                c²
         r2 = c * (17 * 3600 / (1.5 * 10¹¹) )
```

```
Out[65]= 0.06917 AU
```

```
Out[66]= 122.4 AU
```

So star S2 is in no immediate danger.

Problem 4-1 Ramsauer Effect

A 1-eV electron encounters an 8.4 eV potential well. The well is 2 Angstroms wide. What is the probability of transmission? What is the probability the electron is reflected?

```
Plot[{1, -8.4 * UnitStep[x - 1.1] + 8.4 * UnitStep[x - 3.1]},
  {x, -3, 4.1}, PlotRange → {-8.5, 4.1}, Axes → None,
  Prolog → {Text["1 eV", {3.78, 2.45}], Text["-8.4 eV", {3.7, -7.5}]}]
```

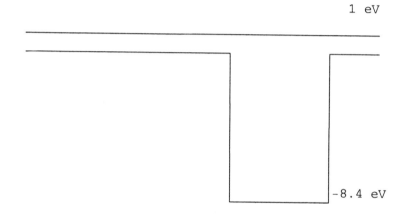

The potential well near a noble-gas atom for 1-eV electron beam

The Quantum wavefunctions in the three regions are :

$$\psi_1 = e^{ikx} + r e^{-ikx} \qquad \text{where } k = \frac{\sqrt{2 m E}}{\hbar}$$

$$\psi_2 = a e^{i\alpha x} + b e^{-i\alpha x} \qquad \text{where } \alpha = \frac{\sqrt{2 m (E + |V|)}}{\hbar}$$

$$\psi_3 = t e^{ikx}$$

```
In[122]:=
    k = √(2 * 9.1 * 10^-31 * 1 * 1.6 * 10^-19) / (6.626 * 10^-34 / (2 π)) / 10^10

    α = √(2 * 9.1 * 10^-31 * 9.4 * 1.6 * 10^-19) / (6.626 * 10^-34 / (2 π)) / 10^10

Out[122]=  k = 0.51171

Out[123]=  α = 1.56887
```

```
Clear[a, b, r, t]
k = .51171; α = 1.56887; L = 2;
sol = NSolve[
   {1 + r == a + b, a * Exp[I * α * L] + b * Exp[-I * α * L] == t * Exp[I * k * L],
    k - k * r == α * a - α * b, α * a * Exp[I * α * L] - α * b * Exp[-I * α * L] ==
      k * t * Exp[I * k * L]}, {a, b, r, t}]
R = Abs[r] ^ 2 /. sol
T = Abs[t] ^ 2 /. sol
psi = Re[Exp[I * k * x] + r * Exp[-I * k * x]] /. sol;
bar = Re[a * Exp[I * α * x] + b * Exp[-I * α * x]] /. sol;
tran = Re[t * Exp[I * k * x]] /. sol;
Plot[psi * If[x < 0, 1, 0] + bar * (UnitStep[x] - UnitStep[x - 2]) +
   tran * If[x > 2, 1, 0], {x, -10, 10}]
```

Out[114]=
$$\{\{a \to 0.663071 - 0.00177808\,i,\ b \to 0.336895 - 0.00349941\,i,$$
$$r \to -0.0000344849 - 0.0052775\,i,\ t \to -0.514851 + 0.857263\,i\}\}$$

Reflection

Out[115]= {0.0000278531}

Transmission

Out[116]= {0.999972}

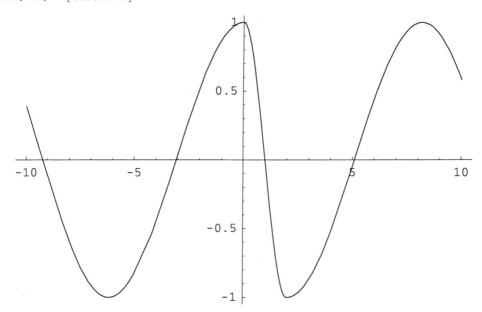

The real part of the wavefunction. Virtually 100% of the incident electron beam is transmitted.

Problem 4-2 Intermediate Quantum Step

We can get 100 % transmission across a potential step if we put in an intermediate quantum step of the appropriate height V and width L. For a 7-eV electron beam impinging on a 5-eV potential step, how high and how wide does the intermediate step need to be? Express the answer in eV and Angstroms.

```
Plot[{7, 2 * (UnitStep[x - 1] - UnitStep[x - 3]), 5 * UnitStep[x - 3]},
   {x, -3, 7}, PlotRange → {-1, 10.1}, Axes → None]
```

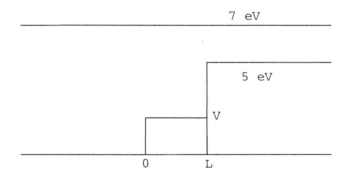

In the three regions,
$$\psi_1 = e^{ikx} \qquad k = \sqrt{2\,m\,(E - 0)}\,\Big/\,\hbar$$
$$\psi_2 = a e^{i\beta x} + b e^{-i\beta x} \qquad \beta = \sqrt{2\,m\,(E - V)}\,\Big/\,\hbar$$
$$\psi_3 = t\,e^{i\gamma x} \qquad \gamma = \sqrt{2\,m\,(E - 5)}\,\Big/\,\hbar$$

Note that the only places where two different wavefunctions and their derivatives will match is at a maximum or minimum, where $\psi'(x) = 0$, or where $\psi(x) = 0$. The distance in phase between these matching points is a multiple of $\pi/2$, or, the minimum width of the step potential is found from $\beta L = \pi/2$. Then, matching wavefunctions and their derivatives at $x = 0$, and $x = L$, we will have 5 equations in 5 unknowns. We will solve for $\{a, b, \beta, L, \text{ and } t\}$ and then check our answer by finding the transmitted quantum flux, $T = (\gamma/k)t^2$. If this is 1, then the values of β and L are correct.

$In[1]:= \quad k = \sqrt{2 * 9.1 * 10^{-31} * 7 * 1.6 * 10^{-19}}\,\Big/\,(6.626 * 10^{-34} \,/\, (2\,\pi))\,\Big/\,10^{10}$

$\gamma = \sqrt{2 * 9.1 * 10^{-31} * (7 - 5) * 1.6 * 10^{-19}}\,\Big/\,(6.626 * 10^{-34} \,/\, (2\,\pi))\,\Big/\,10^{10}$

$Out[1]= \qquad k=1.35386 \qquad\qquad \gamma = 0.723668$

```
In[195]:= Clear[a, b, β, L, t]; k = 1.35386; γ = .723668;
          (*Note 5 equations in 5 unknowns*)
          sol = FindRoot[{1 == a + b,
              a * Exp[I * β * L] + b * Exp[-I * β * L] == t * Exp[I * γ * L],
              k == β * a - β * b, β * a * Exp[I * β * L] - β * b * Exp[-I * β * L] ==
              γ * t * Exp[I * γ * L], β * L == π / 2},
              {β, 1}, {L, 1}, {a, 1}, {b, 1}, {t, 1}] // Chop

          T = (γ / k) * Abs[t] ^ 2 /. sol
          psi = Re[Exp[I * k * x]] /. sol;
          bar = Re[a * Exp[I * β * x] + b * Exp[-I * β * x]] /. sol;
          tran = Re[t * Exp[I * γ * x]] /. sol;
          Plot[psi * If[x < 0, 1, 0] + bar * (UnitStep[x] - UnitStep[x - 2]) +
              tran * If[x > 2, 1, 0], {x, -10, 10}]
```

Out[198]= {β → 0.989821, L → 1.58695, a → 1.18389,
          b → -0.183891, t → 1.24758 + 0.560688 i}

Out[199]= T = 1.

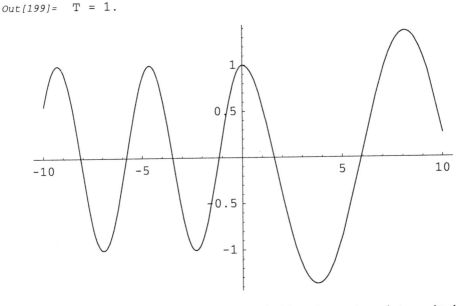

The real part of the wavefunction. 100% of the incident electron beam is transmitted.

```
Clear[V]; L = L /. sol; β = β /. sol; E1 = 7;
soll = Solve[β ==
```

$$\sqrt{2 * 9.1 * 10^{-31} * (7 - V) * 1.6 * 10^{-19}} \Big/ (6.626 * 10^{-34} / (2 \pi)) \Big/ 10^{10}, V\Big]$$

{{V → 3.25834}}

```
In[3]:= Print["L=", L, "A   β=", β, "   V=", V /. soll, "eV"]
```

L =1.58695 A   β =0.989821   V =3.25834 eV

Problem 4-3 Energy Eigenvalues

An electron is trapped in a quantum well. If this is a square well of depth 5 eV and width 2 Angstroms, what is the ground state energy of the electron above the bottom of the well?

The ground-state wavefunction for the electron will be a Cosine wave, because this allows a maximum probability at the center of the well. Notice that it is necessary to match the Cosine wave in the well with an exponentially decreasing wavefunction outside the well.

```
In[2]:= Plot[{-5 * UnitStep[x - 1.1] + 5 * UnitStep[x - 3.1]},
         {x, -3, 4}, PlotRange → {-6.3, 3.1}, Axes → None]
```

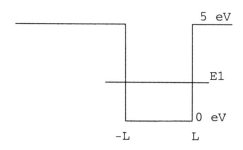

Electron in a Quantum Well

For the ground-state wavefunction, $\psi_2 = A \cos \beta x$ in the well,

and $\psi_3 = C e^{-kx}$ outside the well.

If we set $\psi_2 = \psi_3$ and their derivatives equal at $x = L$

$$A \cos \beta L = C e^{-kL}$$
$$-\beta A \sin \beta L = -k C e^{-kL}$$

Then, $\beta \tan \beta L = k$

Where $\beta = \dfrac{\sqrt{2mE}}{\hbar}$ and $k = \dfrac{\sqrt{2m(V-E)}}{\hbar}$

The above equation may easily be solved in *Mathematica*, however let us first make sure E and V are in eV, and L is in Angstrom units.

$$\beta \, Tan \, \beta \, L = k$$

$$Tan \, \beta \, L = \frac{k}{\beta}$$

$$Tan \left(\frac{\sqrt{2\,m\,E}\,L}{\hbar} \right) = \frac{\sqrt{V - E}}{\sqrt{E}}$$

or $\quad Tan \left(\gamma \, \sqrt{E} \right) = \dfrac{\sqrt{V - E}}{\sqrt{E}}$

where $\quad \gamma = \dfrac{\sqrt{2 * 9.1 \times 10^{-31} * 1.6 \times 10^{-19}}}{(6.626 \times 10^{-34} / 2\,\pi)} * 1 \times 10^{-10} = .5117 \, eV^{-1/2}$

$In[19]:= \quad \gamma = \dfrac{\sqrt{2 * 9.1 * 10^{-31} * 1.6 * 10^{-19}}}{(6.626 * 10^{-34} / (2\,\pi))} * (1 \times 10^{-10})$

$Out[19]= \quad 0.51171$

$In[14]:=$ **Clear[V]; V = 5.0; γ = .51171;**

$\quad$ **sol = FindRoot** $\left[\mathbf{Tan} \left[\gamma * \sqrt{\mathbf{E1}} \right] \; == \; \dfrac{\sqrt{(\mathbf{V - E1})}}{\sqrt{\mathbf{E1}}}, \; \{\mathbf{E1, 1}\} \right]$

$Out[16]= \quad \{E1 \to 2.43473\}$

The ground state energy of the electron is 2.434 eV above the bottom of the well.

A question that should be asked is, whether there are any higher energy eigenstates in the 5-eV square well. The next-higher wavefunction that can fit into the well is $\psi = Sin \, \beta x$. For this wavefunction to match the decaying exponential at the well wall, $\beta x > \pi/2$. (The wavefunction in the well must have a downward slope when it meets the wall).

So for there to be a higher energy eigenstate, $\beta L > \pi/2$, or, for L = 1 A,

$.5117 \, \sqrt{E} > \pi/2$, and E2 > 9.42 eV.

$In[20]:=$ **Solve** $\left[.5117 \, \sqrt{\mathbf{E2}} \; == \; \pi \, / \, 2, \; \mathbf{E2} \right]$

$Out[20]= \quad \{\{E2 \to 9.42343\}\}$

There are no higher energy eigenstates in this 5-eV deep, 2A-wide well.

Problem 4-4 Cooling Concrete in Dams

Given a concrete slab with the following thermal properties, thermal conductivity k = 1.5 BTU/hr/ft²·°F, density ρ = 150 lb/ ft, specific heat C = 0.25 BTU/ °F·lb

What is the temperature in the center of the slab after 40 days?

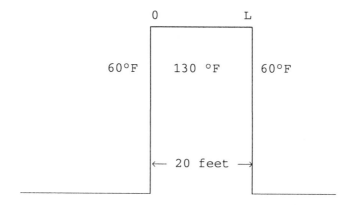

We will use *Mathematica* to numerically solve the PDE

$$\frac{\partial T}{\partial t} = \alpha \frac{\partial^2 T}{\partial x^2}$$

with   (T at x = 0 and x = L) = 0   and T (x, 0) = 70

Where T is the temperature in the slab above the constant outside temperature. The plan is to solve this equation numerically with **NDSolve** and then compare the answer with the exact solution (which for the center of the slab is

$$T\left(\frac{L}{2}\right) = \frac{4 * 70}{\pi} \sum_{k=0}^{10} \frac{(-1)^k}{(2k+1)} * \text{Exp}\left[-\frac{\alpha (2k+1)^2}{(L/\pi)^2} t\right]$$

The diffusivity constant $\alpha = \frac{k}{\rho C}$ = 0.04 ft²/ hr =0.96 ft²/ day with L= 20 ft.

Now, programming*Mathematica* to solve the PDE with the appropriate initial and boundary conditions,

```
In[1]:=  L = 20; α = .96; eq4 = {D[u[x, t], t] - .96 * D[u[x, t], x, x] == 0,
           u[x, 0] == Which[x ≤ 0, 0, 0 < x < 20, 70, x ≥ 20, 0],
           u[0, t] == 0, u[20, t] == 0};
         sol4 = NDSolve[eq4, u[x, t], {x, 0, 20}, {t, 0, 40.0}];
         Plot3D[Evaluate[u[x, t] /. sol4[[1]], {t, 0, 40.0}, {x, 0, 20}]];
```

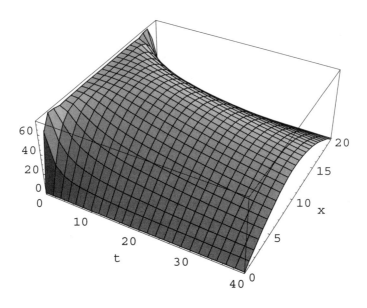

The gradual cooling of the concrete slab.

```
Table[Evaluate[

      {t, u[x, t], 4 * 70 / π  Σ(k=0 to 10) (-1)^k / (2 k + 1) * Exp[- α (2 k + 1)² / (L / π)² t]}] /. sol4[[1]] /.

      x → 10, {t, 2.0, 40.0, 2.0}]] // TableForm
```

t (da)	T (x = 10)	T (Exact)
2.	70.0021	70.
4.	69.9683	69.9569
6.	69.5539	69.5497
8.	68.4981	68.4986
10.	66.851	66.853
12.	64.7889	64.7891
14.	62.4749	62.4741
16.	60.0307	60.0324
18.	57.5485	57.5489
20.	55.0783	55.0785
22.	52.6557	52.6558
24.	50.3008	50.3015
26.	48.0267	48.0277
28.	45.8394	45.8404
30.	43.7413	43.7423
32.	41.7318	41.7333
34.	39.8108	39.812
36.	37.9752	37.9763
38.	36.2222	36.2233
40.	34.549	34.55

The agreement between the numerical solution of *Mathematica* and the exact solution is remarkable. If one desired greater accuracy in the numerical solution, then the command **MaxSteps → 20000** could be added at the end of the **NDSolve** command line.

The final result, with the outside temperature maintained at 60 °F for 40 days and 40 nights, is the center temperature in the slab is a toasty 94.5 °F. Such a temperature differential would cause thermal stresses in the dam, therefore cooling pipes are installed within the concrete to allow the slabs to cool at a uniform rate from center to surface.

For a most interesting description of the methods employed, see Clarence Rawhouser's article "Cooling the Concrete in Grand-Coulee Dam" in <u>Mechanical Engineering</u> (1940) vol. 62, pages 715-718.

For readers who are mathematicians, note that *Mathematica* program only summed the exact solution to a maximum k = 10. This is more than is necessary. A maximum value of k = 5 is entirely satisfactory due to the rapid convergence of the exponential term.

As an additional example of the capabilities of *Mathematica*, notice that the program can evaluate an infinite series

$$In[77]:= \quad \frac{4.0 * 70}{\pi} \sum_{k=0}^{\infty} \frac{(-1)^k}{(2 k + 1)}$$

$$Out[77]= \quad 70.$$

Problem 4-5 Heat Transfer in Steel

<u>Temperature input into a semi-infinite block of steel</u>

A thick steel slab at 550 °F has its surface suddenly cooled to 100 °F. How long before the temperature at 1-inch depth reaches 200 °F ?

The Partial Differential Equation for heat transfer in steel is

$$\frac{\partial u}{\partial t} = \alpha \frac{\partial^2 u}{\partial x^2} \quad \text{where} \quad \alpha = 0.45 \frac{ft^2}{hr} = 0.0075 \frac{ft^2}{min}$$

Initial conditions

  u[x,0] == 550   the block is originally at 550 °F throughout

  u[0,t] == (100+ 450*Exp[-1000*t])

The latter condition is chosen so that the surface is originally at 550 °F, and is then rapidly cooled to 100 °F. In this way, the initial conditions at x=0 are not inconsistent.

Boundary conditions are **D[u[x,t],x]/.x→1) == 0,** so that the spatial derivative vanishes at large x. Since we are looking at a depth of 1-inch, x→1 foot will suffice. The *Mathematica* solution follows:

```
In[47]:=  eq3 = {D[u[x, t], t] - (0.45 / 60) * D[u[x, t], x, x] == 0,
          u[x, 0] == 550, u[0, t] == (100 + 450 * Exp[-1000 * t]),
          (D[u[x, t], x] /. x → 1) == 0};
       sol1 = NDSolve[eq3, u[x, t], {x, 0, 1}, {t, 0, 6}];
       Plot3D[Evaluate[u[x, t] /. sol1[[1]], {t, 0, 6}, {x, 0, 1}]]
```

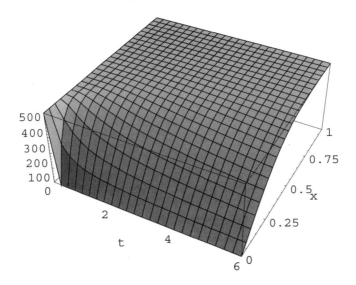

Shown above is the temperature variation in the steel block with u(0 , t) = 100 for t > 0. At a depth of 0.083 ft (1 inch) it should take 5.765 minutes for the temperature to fall from 550 °F to 200 °F. (See Table)

```
Clear@f; f[x_, t_] = u[x, t] /. soll[[1]];
TableForm[Transpose@Prepend[
   Transpose@Table[f[x, t], {t, 0, 6, .25}, {x, 0, 0.415, 0.083}],
     Table[StyleForm[ToString@t, FontWeight → "Bold"], {t, 0, 6, .25}]],
  TableHeadings -> {None, Prepend[
     Table[StyleForm["x(" <> ToString@x <> ")", FontWeight → "Bold"],
       {x, 0, 0.415, 0.083}], StyleForm["t(min)", FontWeight → "Bold"]]}]
```

	0 "	1 "	2"	3 "
t(min)	**x(0)**	**x(0.083)**	**x(0.166)**	**x(0.249)**
0	550.	550.	550.	550.
0.25	100.076	467.188	545.933	550.133
0.5	100.076	402.375	524.115	548.097
0.75	100.076	358.574	497.414	541.203
1.	100.076	328.188	471.965	530.931
1.25	100.076	306.187	449.425	518.995
1.5	100.076	289.526	429.886	506.552
1.75	100.076	276.412	412.995	494.261
2.	100.076	265.745	398.316	482.464
2.25	100.076	256.845	385.446	471.311
2.5	100.076	249.262	374.072	460.841
2.75	100.076	242.694	363.935	451.049
3.	100.076	236.935	354.828	441.904
3.25	100.076	231.824	346.592	433.358
3.5	100.076	227.246	339.101	425.362
3.75	100.076	223.114	332.244	417.87
4.	100.076	219.358	325.939	410.836
4.25	100.076	215.925	320.118	404.222
4.5	100.076	212.781	314.728	398.001
4.75	100.076	209.879	309.714	392.13
5.	100.076	207.189	305.033	386.579
5.25	100.076	204.687	300.652	381.323
5.5	100.076	202.355	296.541	376.338
5.75	100.076	**200.171**	292.669	371.602
6.	100.076	198.123	289.015	367.095

An EXACT solution follows: In either the numerical or Exact case, it takes 5.76 minutes for the temperature at $x = 0.083$ ft to fall from 550 °F to 200 °F.

```
In[13]:= α = 0.45 / 60 ; x1 = .083 ; (*time T1 in minutes *)
         NSolve[{ 200 - (100)
                  ───────────  == Erf[z] , z ==    x1
                  550 - (100)                    ─────────── }, {z, t1}]
                                                 2 √(α * t1)

Out[14]= {{t1 → 5.76635, z → 0.199557}}
```

Problem 5-1 Space Shuttle Phase 2

The second phase of the Space Shuttle launch carries the Shuttle up to the point of orbit insertion. The main engine delivers 5×10^6 N of thrust by burning 1400 kg of liquid H_2 and O_2 per second.

If we take into account the *curvature of the Earth*, we must rewrite the Equations of Motion of the Space Shuttle such that gravitational force decreases with the actual height h of the Shuttle, and is measured toward the center of the Earth. We still measure x and y from the launch point and the height of the Shuttle above the surface (see diagram) is $h = \sqrt{(R+y)^2 + x^2} - R$.

$$my'' = (T-D)\cos(.84 + \epsilon t) - mg_0 \left(\frac{R}{R+h}\right)^2 \cos\lambda \quad y_0 = 47660 \quad y'_0 = 951 \quad \epsilon = .001$$

$$mx'' = (T-D)\sin(.84 + \epsilon t) - mg_0 \left(\frac{R}{R+h}\right)^2 \sin\lambda \quad x_0 = 38620 \quad x'_0 = 1105$$

We may now program these equations into *Mathematica*,

```
m0 = 690000; T = 5 * 10^6; α = 1400; ε = .001; g0 = 9.8; R = 6.4 * 10^6;
m[t] = m0 - α * t; r[t] = √(R+y[t])^2 + x[t]^2; g[t] = g0 * R^2 / r[t]^2;
sol = NDSolve[{
    m[t] * y''[t] == T * Cos[.84 + ε * t] - m[t] * g[t] * (R+y[t]) / r[t],
    m[t] * x''[t] == T * Sin[.84 + ε * t] - m[t] * g[t] * (x[t]) / r[t],
    y[0] == 47660, x[0] == 38620, y'[0] == 951,
    x'[0] == 1105}, {x, y}, {t, 0, 380}]
```

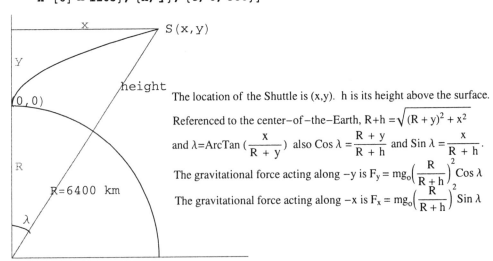

The location of the Shuttle is (x,y). h is its height above the surface.

Referenced to the center-of-the-Earth, $R+h = \sqrt{(R+y)^2 + x^2}$

and $\lambda = \text{ArcTan}\left(\frac{x}{R+y}\right)$ also $\cos\lambda = \frac{R+y}{R+h}$ and $\sin\lambda = \frac{x}{R+h}$.

The gravitational force acting along $-y$ is $F_y = mg_0 \left(\frac{R}{R+h}\right)^2 \cos\lambda$

The gravitational force acting along $-x$ is $F_x = mg_0 \left(\frac{R}{R+h}\right)^2 \sin\lambda$

Location of the Shuttle with respect to the center-of-the-Earth

```
In[33]:=
    InterpFunc1 = x /. sol[[1]]; InterpFunc2 = y /. sol[[1]];
    InterpFunc3 = x' /. sol[[1]]; InterpFunc4 = y' /. sol[[1]];
    height[t_] = √((R + InterpFunc2[t])^2 + InterpFunc1[t]^2) - R;
    tbl = Table[{InterpFunc1[t] / 1000, height[t] / 1000}, {t, 0, 380, 10}];
    ListPlot[tbl, Prolog → AbsolutePointSize[4], AspectRatio → 0.65,
        Epilog → {Text["Distance Downrange(km)", {875, 60}],
        Text["Height of Shuttle(km)", {158, 226}]}];
```

The trajectory of the Space Shuttle from $t = 120$ seconds to $t = 500$ s, just before main fuel-tank separation. Notice that as the Shuttle approaches the end of the second-stage that it begins to gain height. This is because the velocity of the Shuttle is basically along x and the curve of the Earth is falling away beneath the Shuttle.

If we **Table** the results, we will find the height achieved is h = approx 230 km, and the velocity after the second-phase burn is approx 5500 m/s.

```
In[49]:= Table[
            {t, Chop[InterpFunc2[t] / 1000], Chop[InterpFunc1[t] / 1000],
            height[t] / 1000,
            Chop[√(InterpFunc3[t]^2 + InterpFunc4[t]^2)]},
            {t, 0, 380, 10}] // TableForm
```

The (x,y) coordinates and velocity for Space Shuttle second-phase flight

t (s)	y (km)	x (km)	height (km)	v (m/s)
0	47.66	38.62	47.775	1457.88
10	56.930	49.939	57.123	1469.44
20	65.726	61.806	66.021	1486.04
30	74.054	74.237	74.480	1507.66
40	81.923	87.247	82.510	1534.25
50	89.338	100.85	90.122	1565.73
60	96.309	115.06	97.328	1602.01
70	102.84	129.91	104.14	1642.97
80	108.94	145.40	110.56	1688.5
90	114.62	161.57	116.62	1738.48
100	119.88	178.41	122.33	1792.8
110	124.74	195.97	127.69	1851.38
120	129.20	214.26	132.72	1914.11
130	133.27	233.30	137.43	1980.93
140	136.96	253.12	141.85	2051.81
150	140.27	273.74	145.99	2126.7
160	143.22	295.20	149.87	2205.62
170	145.81	317.52	153.51	2288.58
180	148.06	340.73	156.92	2375.64
190	149.97	364.86	160.12	2466.88
200	151.56	389.95	163.15	2562.42
210	152.84	416.04	166.03	2662.4
220	153.81	443.17	168.78	2767.01
230	154.50	471.37	171.43	2876.47
240	154.91	500.70	174.01	2991.06
250	155.06	531.21	176.55	3111.08
260	154.97	562.94	179.10	3236.93
270	154.65	595.96	181.69	3369.04
280	154.12	630.34	184.36	3507.93
290	153.4	666.14	187.16	3654.22
300	152.59	703.43	190.15	3808.61
310	151.43	742.31	193.39	3971.96
320	150.38	782.87	196.93	4145.27
330	149.02	825.22	200.85	4329.72
340	147.78	869.47	205.24	4526.77
350	146.42	915.77	210.18	4738.16
360	145.16	964.26	215.78	4966.01
370	143.88	1015.1	222.16	5213.
380	142.72	1068.5	229.46	5482.48

After the main fuel tank is jettisoned, the 110,000 kg Shuttle is moving at 5500 m/s on a downrange trajectory. Its height is approx 230 km. At this point, the orbit maneuvering system takes over with two 27,000 N thrust engines and increases the speed to approx 7700 m/s which is appropriate for a 400-km orbit above the Earth.

Problem 5-2 B-Z Chemical Reaction

I. Solve the B-Z equations for $x = [HBrO_2]$, $y = [Br^-]$, and $z = [Ce^{+4}]$

$$\epsilon x' = x + y - q x^2 - xy$$

$$7y' = -y + 2hz - xy$$

$$p z' = x - z$$

where $\epsilon = 0.21$, $p = 14$, $q = 0.006$, $h = 0.75$, $x(0) = 100$, $y(0) = 1$, $z(0) = 10$

```
ε = .21; p = 14; q = .006; h = .75;
sol = NDSolve[{ε * x'[t] == x[t] + y[t] - q * x[t]^2 - x[t] * y[t],
    7 y'[t] == -y[t] - x[t] * y[t] + 2 h * z[t], p * z'[t] == x[t] - z[t],
    x[0] == 100, y[0] == 1, z[0] == 10}, {x, y, z}, {t, 0, 100}]
InterpFunc1 = x /. sol[[1]]; InterpFunc2 = y /. sol[[1]];
InterpFunc3 = z /. sol[[1]];
Chop[Table[{t, InterpFunc1[t], InterpFunc2[t], InterpFunc3[t]},
    {t, 0, 80, 2}] // TableForm]
Plot[Evaluate[{x[t] / 4, y[t], z[t]} /. sol],
  {t, 0, 100}, PlotRange → {0, 45},
  PlotStyle → {{RGBColor[0, 1, 0]},
    {RGBColor[1, 0, 0]}, {RGBColor[0, 0, 1]}}]
```

Again, as in Section 5-2, *Mathematica* handles stiff differential equations with no problem.

Let's Table the values to get an accurate read on the time of oscillation. Scanning the Table, we find maximum x at 2 and 56 s, maximum y at 12 and 66 s, and maximum z at 4 and 58 s. The time of oscillation of x, y, and z is therefore 54 seconds.

t(s)	x	y	z
0	100.	1.	10.
2	**122.521**	0.299541	25.2651
4	71.9039	0.687823	**35.1345**
6	1.16892	7.06071	33.7346
8	1.07679	13.9677	29.3908
10	1.06385	16.5633	25.6205
12	1.06249	**16.8911**	22.3513
14	1.06579	16.0882	19.5174
16	1.07202	14.7776	17.0615
18	1.08069	13.294	14.9335
20	1.09171	11.8114	13.0901
22	1.10527	10.4148	11.4938
24	1.12168	9.1409	10.1119
26	1.1414	8.00059	8.91642
28	1.16509	6.99132	7.88296
30	1.1936	6.10419	6.99055
32	1.22806	5.32765	6.22112
34	1.27003	4.64947	5.5592
36	1.32164	4.05772	4.99161
38	1.386	3.54121	4.50728
40	1.46773	3.08959	4.09711
42	1.57423	2.69316	3.75401
44	1.71838	2.34255	3.47313
46	1.9255	2.02771	3.25273
48	2.25613	1.73545	3.0966
50	2.91417	1.44012	3.02339
52	5.37809	1.04398	3.12871
54	105.093	0.0963774	6.47907
56	**127.011**	0.27174	23.7781
58	79.1861	0.620734	**34.5446**
60	1.21035	5.98529	34.2246
62	1.07966	13.5059	29.817
64	1.06441	16.431	25.9901
66	1.06235	**16.928**	22.6717
68	1.06528	16.2059	19.7952
70	1.07125	14.9282	17.3022
72	1.07967	13.4519	15.1421
74	1.09044	11.9638	13.2708
76	1.10372	10.5557	11.6502
78	1.11981	9.26806	10.2473
80	1.13917	8.11368	9.03348

II.   Solve the B-Z equations

$$\epsilon \, x' = \, x + y - q \, x^2 - xy$$

$$7 \, y' = \, -y + 2hz - xy$$

$$p \, z' = \, x - z$$

where $\epsilon = 0.21$,   p=14,   q= 0.006,   h =0.75,   x(0) = 17,   y(0) = 3,   z(0) = 5

```
ε = .21; p = 14; q = .006; h = .75;
sol = NDSolve[{ε*x'[t] == x[t] + y[t] - q*x[t]^2 - x[t]*y[t],
    7y'[t] == -y[t] - x[t]*y[t] + 2h*z[t], p*z'[t] == x[t] - z[t],
    x[0] == 17, y[0] == 3, z[0] == 5}, {x, y, z}, {t, 0, 100}]
InterpFunc1 = x /. sol[[1]]; InterpFunc2 = y /. sol[[1]];
InterpFunc3 = z /. sol[[1]];
Chop[Table[{t, InterpFunc1[t], InterpFunc2[t], InterpFunc3[t]},
    {t, 0, 90, 2}] // TableForm]
Plot[Evaluate[{x[t]/4, y[t], z[t]} /. sol],
  {t, 0, 100}, PlotRange → {0, 45},
  Prolog → {Text["x/4", {15, 30}], Text["y", {29, 14}],
    Text["t(s)", {96.5, 2.3}], Text["Z", {31.2, 26}]},
  PlotStyle →
    {{RGBColor[0, 1, 0]}, {RGBColor[1, 0, 0]}, {RGBColor[0, 0, 1]}}]
```

Again, as in Section 5.2, *Mathematica* handles the stiff-differential equations with no problem. The effect of varying the reactants is to delay the onset of the oscillations. x, y, and z have to build up to a certain level before the color-changes begin.

Let's Table the values to get an accurate read on the time of oscillation.

t(s)	x	y	z
0	17.	3.	5.
2	1.65736	2.50447	4.67582
4	1.64866	2.51232	4.27255
6	1.70065	2.39096	3.92624
8	1.80322	2.20369	3.63636
10	1.96856	1.98605	3.40276
12	2.23804	1.75373	3.22862
14	2.74323	1.50346	3.12704
16	4.18261	1.18636	3.15243
18	28.8258	0.375095	4.02654
20	**138.115**	0.197779	18.876
22	96.9563	0.478375	32.1197
24	7.25794	2.49496	**35.6603**
26	1.09234	11.8139	31.1106
28	1.06688	15.8651	27.1124
30	1.06229	**16.9474**	23.6447
32	1.06401	16.511	20.6386
34	1.06913	15.3556	18.0331
36	1.07682	13.9146	15.7753
38	1.08686	12.417	13.8193
40	1.09935	10.9781	12.1251
42	1.11453	9.65103	10.6584
44	1.13282	8.45526	9.38907
46	1.15478	7.39267	8.29143
48	1.18118	6.4564	7.34313
50	1.21302	5.63566	6.52494
52	1.25167	4.91834	5.82036
54	1.29899	4.2923	5.21528
56	1.35763	3.74603	4.69779
58	1.43149	3.26882	4.25797
60	1.5266	2.85075	3.88791
62	1.6531	2.48237	3.58176
64	1.82977	2.15408	3.33633
66	2.0976	1.85452	3.15263
68	2.57319	1.56537	3.04098
70	3.81791	1.23611	3.04638
72	18.9211	0.518451	3.64735
74	**141.401**	0.174242	17.0551
76	102.118	0.440834	31.1047
78	18.9609	1.71081	**36.0066**
80	1.09873	11.128	31.5625
82	1.06805	15.6137	27.5044
84	1.06238	**16.9237**	23.9846
86	1.06364	16.6004	20.9332
88	1.06847	15.4952	18.2884

Judging from peaks in x, y, and z the period of oscillation is about 54 seconds.

Problem 5-3 Equations of Winemaking

We will solve the equations of winemaking

$$x' = (a - bx - cy_*) \, x \qquad x = \text{\# yeast}$$
$$y' = k \, x \qquad\qquad y = \text{alcohol per cent}$$
$$z' = -\gamma \, x \qquad\qquad z = \text{remaining sugar (grams)}$$

with initial conditions $x(0) = 100$, $y(0) = 0$, $z(0) = 1140$

$a = 0.125$, $b = a/10^7$, $c = a/10$, $y_* = y - 4$ (before $y = 4$, $c = 0$)

$\gamma = 10^{11}$ sugar molecules/second/yeast $= \frac{180*10^{11}}{6*10^{23}}$ grams/second/yeast.

$k = 3.4 \times 10^{-13}$ percent/ second/ yeast

We will measure time in hours. Then, programming the three equations into *Mathematica*, we will first plot out the yeast population x for 10 days, and then **Table** the values of x , y , and z for 70 days.

```
a = 1 / 8.; (*e-folding time of 8 hr*) k = 3600 * 3.4 * (10^-13);
b = a / (10^7);  (*max yeast population of 10^7 *)
c = a / 10; d = 3600 * 10^11 * 180 / (6 * 10^23);
sol1 =
 NDSolve[{x'[t] == (a - b * x[t] - c * If[y[t] < 4, 0, 1] * (y[t] - 4)) * x[t],
   y'[t] == k * x[t], z'[t] == -d * x[t], x[0] == 100,
   y[0] == 0, z[0] == 1140}, {x, y, z}, {t, 0, 1680}]
InterpFunc1 = x /. sol1[[1]]; InterpFunc2 = y /. sol1[[1]];
InterpFunc3 = z /. sol1[[1]];
ListPlot[Table[{t / 24, InterpFunc1[t]}, {t, 0, 240, 2}]]
```

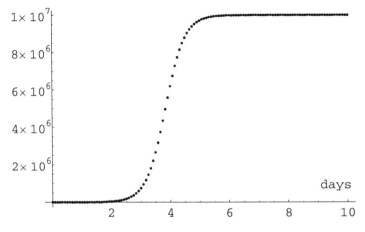

The yeast population reaches its maximum in about 6 days.

```
Plot[Evaluate[{x[t] / 10^6, y[t]} /. sol1, {t, 0, 1200}],
  Prolog → {Text["yeast population", {310, 9.0}], Text["(millions)",
    {330, 8.3}], Text["Alcohol percent", {925, 8.0}],
  Text["Hours", {1145, 0.8}]}]
```

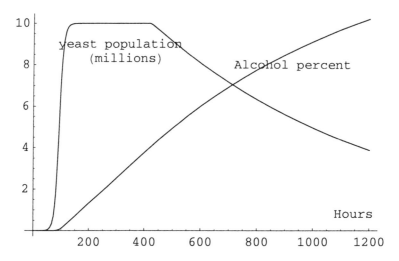

The progress of winemaking. Yeast population and alcohol percentage for the first 50 days.

Let us now **Table** the functions x, y, and z

x = # of yeast, y = Alcohol percent, z = Sugar supply in grams

```
tbl = Table[{t / 24, InterpFunc1[t] / 10^6, Chop[InterpFunc2[t]],
    InterpFunc3[t]}, {t, 0, 1680, 48}] // TableForm
```

Time(da)	x Millions of yeast	y Alcohol %	z Sugar(gm)
0	0.0001	0	1140.
2	0.0401811	0.000393267	1139.97
4	6.19419	0.0945953	1131.65
6	9.98479	0.635363	1083.94
8	9.99996	1.22273	1032.11
10	10.	1.81025	980.272
12	10.	2.39777	928.432
14	10.	2.98529	876.592
16	10.	3.57281	824.752
18	9.91893	4.15985	772.955
20	9.37113	4.72673	722.936
22	8.83648	5.26144	675.755
24	8.33227	5.76565	631.267
26	7.85684	6.24108	589.316
28	7.40853	6.68939	549.76
30	6.98581	7.11211	512.461
32	6.5872	7.51072	477.289
34	6.21134	7.88658	444.125
36	5.85692	8.241	412.853
38	5.52273	8.57519	383.366
40	5.20761	8.89031	355.561
42	4.91046	9.18746	329.342
44	4.63028	9.46764	304.62
46	4.36607	9.73185	281.308
48	4.11695	9.98097	259.326
50	3.88204	10.2159	238.599
52	3.66053	10.4374	219.054
54	3.45166	10.6463	200.624
56	3.25471	10.8432	183.247
58	3.069	11.0289	166.86
60	2.89389	11.204	151.409
62	2.72876	11.3692	136.839
64	2.57306	11.5249	123.101
66	2.42624	11.6717	110.146
68	2.2878	11.8101	97.9309
70	2.15726	11.9407	86.4126

See Cyril J Berry's 1994 book <u>First Steps in Winemaking</u> for all the steps necessary to make a good wine. As noted in Berry's book, different strains of yeast will have different tolerances to the alcohol percentage. Also, the dryness or sweetness of the wine is determined by how long the fermentation process takes. The alcohol percentage in the wine may be gauged at any time by utilizing the hydrometer. Winemaking is an art as well as a science.

Problem 5-4 H-H Nerve Conduction

Let us use the equations of Section 5-4 to find the form of the action potential at 18.5 °C. With α, β, γ, δ, ϵ, and η all multiplied by the temperature factor $3^{(T-6.3)/10} = 3.82$, we will find the conduction velocity.

We will then use the H–H model to show that when a nerve fiber is at 18.5 °C that a 29-mV, 0.5 ms half-sine pulse will elicit an action potential whereas a 28-mV, 0.5 ms half-sine pulse will <u>not</u> produce an action potential.

```
c = .001; a = .0238; ρ = 35.4;

eq3 = { ───── D[u[x, t], x, x] ==
        2 * ρ

    c * D[u[x, t], t] + .036 * n[x, t] ^ 4 * (u[x, t] + 12) +
    .120 * m[x, t] ^ 3 * h[x, t] * (u[x, t] - 115) + .0003 * (u[x, t] - 10.6),
    u[x, 0] == 0, u[0, t] == (60 * Sin[2 π * t] * If[0 < t < 0.5, 1, 0]),
    (D[u[x, t], x] /. x → 3.2) == 0};

eq4 = {D[n[x, t], t] == ─────────────────────────── * (1 - n[x, t]) -
                          .0382 * (10 - u[x, t])
                          Exp[(10 - u[x, t]) / 10] - 1

    .4775 * Exp[- ──────── ] * n[x, t], n[x, 0] == .3177};
                  u[x, t]
                    80

eq5 = {D[m[x, t], t] == ─────────────────────────── * (1 - m[x, t]) -
                          .382 * (25 - u[x, t])
                          Exp[(25 - u[x, t]) / 10] - 1

    15.28 * Exp[- ──────── ] * m[x, t], m[x, 0] == .0529};
                  u[x, t]
                    18

eq6 = {D[h[x, t], t] == .2674 * Exp[- ──────── ] * (1 - h[x, t]) -
                                      u[x, t]
                                        20

        3.82
    ─────────────────────────── * h[x, t], h[x, 0] == .5961};
    Exp[(30 - u[x, t]) / 10] + 1

sol1 = NDSolve[{eq3, eq4, eq5, eq6},
    {u[x, t], n[x, t], m[x, t], h[x, t]}, {x, 0, 3.2}, {t, 0, 3.6}];
Plot3D[Evaluate[u[x, t] /. sol1[[1]], {x, 0, 2.0}, {t, 0, 2}],
    PlotRange → {0, 100}, PlotPoints → 30, AxesLabel → {"x", "t", " "},
    DisplayFunction → $DisplayFunction, ViewPoint → {2.5, -1.5, .95}];
Clear@f; f[x_, t_] = u[x, t] /. sol1[[1]]
Pictures = Table[f[x, t], {x, 0, 2.5, .5}]
Plot[Evaluate[Pictures], {t, 0, 3.0}, Frame → True,
    PlotRange → All, Prolog → { Text["x=0", {.25, 64}],
    Text["1cm", {0.88, 74}], Text["x=2.5cm", {2., 74}]}]
```

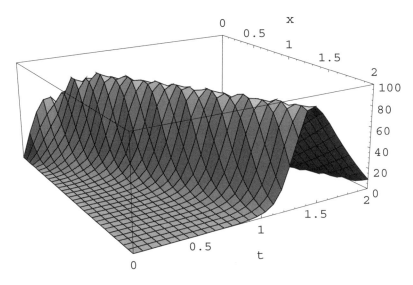

Response of nerve axon to 60-mV input

If we plot the nerve impulse as it is seen at various points along the axon

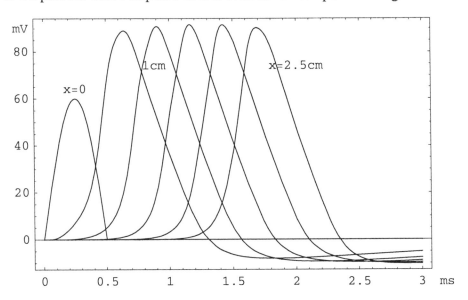

Note that the pulse is "sharper" at 18.5°C -- the conduction velocity is ~ 18.8 m/s

The H–H model shows that the action potential at 18.5 °C is much "sharper" than the impulse at 6.3 °C, and travels much faster. The peak of the action potential passes $x = 1$ cm at 0.9 ms, and passes the 2.5-cm mark at 1.7 ms, then the conduction velocity at 18.5 °C is $v = \frac{1.5 \text{ cm}}{0.8 \text{ ms}} = 18.8$ m/s. This is the same value found by Hodgkin and Huxley.

We may also check the conduction velocity by creating a table of values of voltage versus time and distance. The maximum voltages (89-90 mV) occur at   (x = 1.0 cm, t = 0.9 ms) and (x = 2.5 cm, t = 1.7 ms). Therefore the conduction velocity is $v = \frac{\Delta x}{\Delta t} = \frac{1.5 \text{ cm}}{0.8 \text{ ms}} = 18.8$ m/s

```
TableForm[Transpose@
    Prepend[Transpose@Table[f[x, t], {t, 0, 3, .1}, {x, 0, 2.5, 0.5}],
      Table[StyleForm[ToString@t, FontWeight → "Bold"],
        {t, 0, 3, .1}]] // Chop,
    TableHeadings -> {None, Prepend[
      Table[StyleForm["x(" <> ToString@x <> ")", FontWeight → "Bold"],
        {x, 0, 2.5, 0.5}], StyleForm["t(ms)", FontWeight → "Bold"]]}]
```

t(ms)	x(0)	x(0.5)	x(1.)	x(1.5)	x(2.)	x(2.5)
0	0	0	0	0	0	0
0.1	35.26	0.62	-0.00	-0.00	-0.00	-0.00
0.2	57.06	4.43	0.071	-0.00	-0.00	-0.00
0.3	57.06	11.5	0.572	0.007	-0.00	-0.00
0.4	35.26	28.1	1.889	0.071	0.000	-0.00
0.5	0.00	66.5	5.176	0.277	0.007	-0.00
0.6	0.00	**87.1**	14.09	0.846	0.037	0.000
0.7	0.00	84.8	40.02	2.446	0.126	0.004
0.8	0.00	68.9	75.69	6.835	0.388	0.017
0.9	0.00	51.9	**90.81**	19.97	1.155	0.057
1.	0.00	36.2	80.18	52.71	3.343	0.178
1.1	0.00	21.7	63.51	82.96	9.499	0.545
1.2	0.00	8.87	46.82	**89.37**	28.45	1.634
1.3	0.00	0.39	31.46	75.14	61.18	4.778
1.4	0.00	-4.2	17.02	58.18	**90.70**	13.82
1.5	0.00	-6.4	5.236	41.92	85.12	36.48
1.6	0.00	-7.4	-2.14	26.79	69.97	74.77
1.7	0.00	-7.9	-5.79	12.95	52.98	**90.09**
1.8	0.00	-8.0	-7.64	2.359	37.10	81.84
1.9	0.00	-8.0	-8.61	-3.63	22.34	65.04
2.	0.00	-7.8	-9.09	-6.57	9.231	47.68
2.1	0.00	-7.6	-9.30	-8.12	-0.03	31.88
2.2	0.00	-7.4	-9.34	-8.94	-4.77	17.16
2.3	0.00	-7.1	-9.27	-9.36	-7.23	4.395
2.4	0.00	-6.8	-9.13	-9.55	-8.54	-3.45
2.5	0.00	-6.5	-8.94	-9.59	-9.26	-7.39
2.6	0.00	-6.2	-8.71	-9.55	-9.65	-9.18
2.7	0.00	-5.9	-8.45	-9.44	-9.82	-9.93
2.8	0.00	-5.6	-8.18	-9.29	-9.84	-10.2
2.9	0.00	-5.3	-7.90	-9.10	-9.78	-10.2
3.	0.00	-5.0	-7.60	-8.88	-9.64	-10.1

Now consider the response of the nerve axon to a 29-mV pulse.

```
c = .001; a = .0238; ρ = 35.4;
eq3 = { ──a── D[u[x, t], x, x] ==
        2 * ρ
    c * D[u[x, t], t] + .036 * n[x, t]^4 * (u[x, t] + 12) +
      .120 * m[x, t]^3 * h[x, t] * (u[x, t] - 115) + .0003 * (u[x, t] - 10.6),
    u[x, 0] == 0, u[0, t] == (29 * Sin[2 π * t] * If[0 < t < 0.5, 1, 0]),
    (D[u[x, t], x] /. x → 3.2) == 0};
eq4 = {D[n[x, t], t] == ──.0382 * (10 - u[x, t])── * (1 - n[x, t]) -
                        Exp[(10 - u[x, t]) / 10] - 1
    .4775 * Exp[-u[x, t] / 80] * n[x, t], n[x, 0] == .3177};
eq5 = {D[m[x, t], t] == ──.382 * (25 - u[x, t])── * (1 - m[x, t]) -
                        Exp[(25 - u[x, t]) / 10] - 1
    15.28 * Exp[-u[x, t] / 18] * m[x, t], m[x, 0] == .0529};
eq6 = {D[h[x, t], t] == .2674 * Exp[-u[x, t] / 20] * (1 - h[x, t]) -
            ──────3.82────── * h[x, t], h[x, 0] == .5961};
            Exp[(30 - u[x, t]) / 10] + 1
sol1 = NDSolve[{eq3, eq4, eq5, eq6}, {u[x, t], n[x, t], m[x, t], h[x, t]},
    {x, 0, 3.2}, {t, 0, 3.6}];
Plot3D[Evaluate[u[x, t] /. sol1[[1]]], {x, 0, 1.5}, {t, 0, 2.5},
    PlotRange → {0, 100}, PlotPoints → 30, AxesLabel → {"x", "t", " "}];
Clear@f; f[x_, t_] = u[x, t] /. sol1[[1]]
Pictures = Table[f[x, t], {x, 0, 2.5, .5}]
Plot[Evaluate[Pictures], {t, 0, 3.0}, Frame → True, PlotRange → All]
```

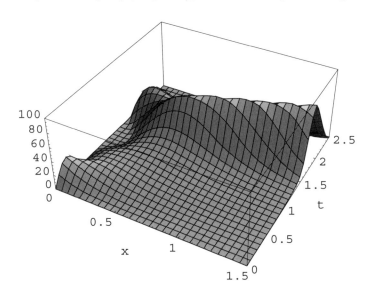

The action potential of the 29-mV input

If we run the above program with a 28-mV input, we get this response

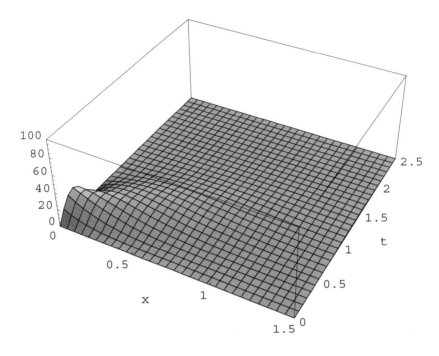

The 28-mV impulse falls away to nothing

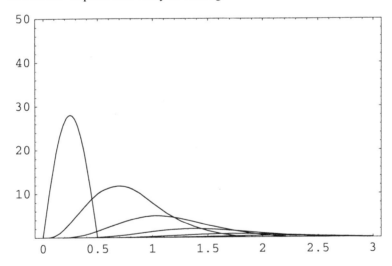

Failure of 28-mV input to produce an action potential

The "all-or-nothing" response of a nerve fiber is essential to keep noise off the line. If the threshold of minimum input is not met, then the nerve does not conduct.

Problem 5-5 Flight to the Stars

In this final problem, we will find the time required to get to the nearest star at 1 g acceleration.

Let us first put all the distances in lightyears, and measure time in years. Thus $c = 1$ ly/y, and $g = 9.81 \text{m/s}^2 = 1.032$ ly/y^2. We shall continue to measure mass in kilograms.

Now, we will carefully find the velocity v of the starship as a function of Earth-time t.

$$\frac{dv}{d\tau} = \frac{d}{dt}\{\gamma\, v\} = g$$

we will use *Mathematica* to find $v[t]$

```
In[18]:=  γ[t_] = (1 - v[t]² / c²)^{-1/2}; c = 1;
          DSolve[{D[γ[t] * v[t], t] == g, v[0] == 0}, v[t], t];
          Simplify[%, {t > 0, g > 0}]
```

$$Out[20]= \left\{\left\{v[t] \to -\frac{g\,t}{\sqrt{1 + g^2\,t^2}}\right\}, \left\{v[t] \to \frac{g\,t}{\sqrt{1 + g^2\,t^2}}\right\}\right\}$$

Now that we have v as a function of acceleration, g, and Earth-time t, we may find the length of time necessary to travel half-way to Alpha-Centauri.

```
In[32]:=  g = 1.032; c = 1.0;  v[t_] = ────g * t──── ;
                                    √(1 + g² t² / c²)
          (* Given v[t], we integrate to time t to find x[t]*)
```

$$x[t\_] = \int_0^t v[t]\, dt$$

$$Out[34]= \left\{\left\{x[t] \to -0.968992 + 0.968992\,\sqrt{1. + 1.06502\,t^2}\right\}\right\}$$

Even more conveniently we may solve the following for the time t at which $x = 2.2$ lightyears

$$\text{Solve}\left[2.2 == \int_0^t v[t]\, dt, t\right]$$

```
{{t→-3.01721},{t→3.01721}}
```

Therefore, it takes 3 years (Earth-time) for the starship to reach midpoint.

At that time, the velocity of the starship is .9515 c.

In[69]:= v[3]

Out[69]= 0.951593

        Plot[{v[t], .95}, {t, 0, 3}, PlotRange → {{-0.3, 3.06}, {0, .99}},
        AxesLabel → {"EarthTime(y)", "Velocity (v/c)"}]

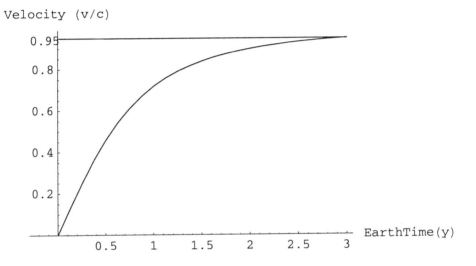

The approach to lightspeed

We have the velocity and distance traveled for the starship in Earthyears. Let us now find the distance traveled and the time for the space-travelers. Let z[t] be the distance the astronauts see as being traveled, and τ[t] be the time elapsed.

In[35]:= $z[t\_] = \int_0^t \frac{v[t]\ dt}{\gamma[t]}$ // FullSimplify; Simplify[%, t > 0]

Out[36]= $z[t] \to 0.030522 + 0.484496\ \mathrm{Log}[0.938946 + 1.\ t^2]$

In[215]:= $\tau[t\_] = \int_0^t \frac{dt}{\gamma[t]}$ // FullSimplify

Out[216]= $\tau[t] \to 0.968992\ \mathrm{ArcSinh}[1.032\ t]$

Note that z[3] = 1.14 lightyears and τ[3]= 1.79 years as seen on the ship.

Therefore the astronauts see the trip to Alpha-Centauri as taking 3.58 years, and a distance of only 2.28 lightyears.

We may **Table** our results and check on the astronauts every 3 months from our Earth-bound observatory.

Table[{t, τ[t], x[t], z[t], v[t], γ[t]}, {t, 0, 3, .25}] // TableForm

Earth-time t (yr)	Ship-time τ (yr)	Distance traveled (our frame) x (ly)	Distance traveled (ship-frame) z (ly)	v/c	γ
0	0	0.	0	0	1
0.25	0.247306	0.031735	0.031222	0.249819	1.0327
0.5	0.480113	0.121396	0.114372	0.458552	1.1252
0.75	0.690152	0.256343	0.227435	0.612077	1.2645
1.	0.875796	0.423468	0.351328	0.718153	1.4370
1.25	1.03909	0.612604	0.474741	0.790341	1.6322
1.5	1.18323	0.81677	0.592387	0.839978	1.8429
1.75	1.31139	1.03137	0.702351	0.874842	2.0643
2.	1.42626	1.25338	0.804336	0.899939	2.2934
2.25	1.53005	1.48079	0.898739	0.918448	2.5281
2.5	1.62455	1.71223	0.986212	0.932411	2.7670
2.75	1.71118	1.94673	1.06746	0.943162	3.0090
3.	1.79106	2.18362	1.14315	0.951593	3.2534

We now have sufficient information to answer the questions posed in part (A).
The trip as seen by the astronauts will take 4 τ + 1 years = 8.2 years
The trip as seen from Earth will take   4 t + 1 years = 13 years.

At this time it is worthwhile checking to be sure we have done everything right by computing the acceleration of the spacecraft, say from t=1.000 y to 1.001 yr and seeing if this matches the correct relativistic formula for measurements made in the *accelerated* frame: $a' = a_x \gamma^3$ (see John David Jackson *Classical Electrodynamics,* or Paul Lorrain and Dale Corson *Electromagnetic Fields and Waves,* for derivation of relativistic formulas).

$$In[39]:= \quad \frac{v[1.001] - v[1.000]}{1.001 - 1.000} * \gamma[1.0005]^3$$

Out[39]= 1.032

This is g in ly/y^2 so the astronauts see the engines providing the necessary acceleration whereas we on Earth would see the spacecraft as accelerating only at 1/3 g.

At t = 3 y Earth-time, or τ = 1.8 y Ship-time the spacecraft is up-to-maximum speed. It is now time to turn the craft through 180° and begin the slow-down to Alpha-Centauri. However, how much matter and antimatter have we burned? This may be computed with the thrust equation, noting that time is in τ

$$- \frac{dm}{d\tau} c = mg \qquad \text{where the starting mass of the ship is M}$$

```
In[43]:= Clear[c, g, v];
         DSolve[{m'[τ] * c == - m [τ] * g, m[0] == M}, m[τ], τ]
```

$$Out[44]= \left\{\left\{m[\tau] \rightarrow e^{-\frac{g\tau}{c}} M\right\}\right\}$$

Thus at the end of the first-stage burn, the mass of the starship is reduced by a factor of **Exp[g*τ/c]**. For g = 1.032 ly/y^2, τ = 1.8 y, c = 1 ly/y,

```
In[42]:= Exp[1.032 * 1.8]
```

```
Out[42]= 6.40834
```

For each stage (and there are four of them) we reduce the mass of the previous stage by a factor of 6.4. Thus if the *returning* spacecraft has a mass of 5000 kg, then the original mass of the starship is

$$M = 6.4^4 * 5000 \text{ kg} = 8.4 \text{ million kg} = 8400 \text{ metric tons}$$

This actually compares favorably with the at-launch mass of the 3000 metric ton Saturn-V rocket, until one realizes that the α-Centauri rocket requires

4200 metric tons of matter <u>and</u>

4200 metric tons of antimatter !

Notes on the Text

1.1 Motion of the Planets about the Sun

A question that naturally arises when solving Newton's equations with numerical methods is "How difficult is it to determine the track of an asteroid ?" Clearly one can find the orbital path of an asteroid if you are given a portion of its orbit that is well-defined. Then the *initial conditions* are known. You may then develop a 3-body problem, using the coordinates of the asteroid, the sun, and the moving Earth. This will also be a 3-dimensional problem, because the Earth and the asteroid, in general, will not be in the same plane. But it will be solvable, by computer, <u>if</u> the starting distances and original velocity vector of the asteroid are well-determined.

Problem 1-3 Solar Sailing

A new material, Kapton, has been developed which is essentially a thin film of micron thickness which can be aluminized with a coating approximately 0.1 microns in thickness. With a weight density of 1 gram per square meter for the sail, solar sailing could become a reality, if a reliable means for packaging and unfurling the sail can be found.

For more on solar sailing, see *Space Sailing* by Jerome Wright (1992), *Solar Sailing* by Colin McInnes (1999), *or* the article by T. C. Tsu (1959) *"Interplanetary Travel by Solar Sail."* American Rocket Society 422-427.

2.1 Damped and Driven Oscillations

In vibration analysis, engineers utilize the *damping ratio* $z = \dfrac{c}{2\sqrt{mk}}$

For $z > 1$, the motion is overdamped,

for $z = 1$, the motion is critically damped,

and for $z < 1$, the motion is underdamped.

Using the *damping ratio,* the frequency of underdamped motion is

$$\omega_d = \omega_o \sqrt{1 - z^2}$$

where $\omega_o = \sqrt{\dfrac{k}{m}}$ and the logarithmic decrement is $\delta = \dfrac{c\,\pi}{m\,\omega_d} = \dfrac{2\,\pi\,z}{\sqrt{1-z^2}}$

2.2 Shock Absorbers

$e^{-\delta}$ is the amount of damping in one cycle.

For new shock absorbers $\delta \simeq 4$, for an older automobile $\delta \simeq 2$.

2.5 Variable Stars

The mathematics of the (Eddington) variable star equations are very interest-ing. For $x^3 \ddot{x} = -Ax + B$ this is an oscillating system as long as $B > Ax_0$. Then at $t = 0$, the stellar envelope will be accelerated towards $+x$. (Pressure exceeds gravity.)

When x gets large enough, then $Ax > B$ and $\ddot{x}$ is negative. (Gravity exceeds pressure). Notice also that when $Ax = B$, the system is passing through its equilibrium point. At that time, $x = B/A$ and putting in the numbers from Section 2.5, $x = 1.50 \times 10^7$ km.

Problem 2-5 Impulse Response

Theoretically, we could model a piano sonata as the response of the piano strings to the impulse functions of the hammers striking the strings.

3.3 The Binary Pulsar

Astronomers find the *eccentricity* of a binary system by measuring the Doppler shift due to the velocity of the stars. The maximum and minimum velocities of the stars are in the ratio $\dfrac{V_P}{V_A} = \sqrt{\dfrac{1 - \epsilon}{1 + \epsilon}}$

3.5 Heat Exchanger

In the engineering literature, the usual method for finding the effectiveness of a heat-exchanger is to evaluate

parallel-flow

$$\epsilon = \frac{1 - \mathrm{Exp}[-NTU*(1+C^*)]}{(1+C^*)} \quad \text{where NTU} = \frac{UA}{(\dot{m}c)_{min}} \quad \text{and} \quad C^* = \frac{(\dot{m}c)_{min}}{(\dot{m}c)_{max}}$$

counter-flow

$$\epsilon = \frac{1 - \mathrm{Exp}[-NTU*(1-C^*)]}{1 - C^* \, \mathrm{Exp}[-NTU*(1-C^*)]} \quad \text{where NTU} = \frac{UA}{(\dot{m}c)_{min}} \quad \text{and} \quad C^* = \frac{(\dot{m}c)_{min}}{(\dot{m}c)_{max}}$$

Problem 3-5 Heat Exchangers

It is amazing, but true, that if we reverse the direction of flow of oil in the parallel-flow heat exchanger, that we will now have a counterflow heat exchanger that will extract 11% more heat energy. This is because of the more uniform thermal gradient in the counterflow system.

4.1 Partial Differential Equations

The PDE for waves on a string is

$$\frac{\partial^2 u}{\partial t^2} = c^2 \frac{\partial^2 u}{\partial x^2}$$

Boundary condition   u (x, t) = 0 at x = 0 and x = L
Initial condition   u (x, 0) = f[x] = 0 for 0 < x < L
            and   D[u (x, 0), t] = g[x]

As an example, take a piano string under tension T and with mass per unit length of ρ. Then with the hammer striking the string at t = 0, with an impulse function of fairly small width, **g(x)** is the initial velocity of the string. See the *Mathematica* file **PDE-Wave** on the CD for the necessary code to solve the Wave Equation.

4.3 PDE – Heat Conduction in the Earth

In their book on Heat Transfer, Incropera and DeWitt treat the case of 20 °C soil subjected to a – 15 °C surface freeze for 60 days. For $\alpha = 0.138 \times 10^{-6}$ m^2/s, *Mathematica* shows that the ground will be frozen (T = 0 °C) to a depth of 0.68 meters. (See **PDE-Incropera** on the CD)

5.2 The B–Z Chemical Reaction

According to John Tyson (1976, see References), the B-Z reaction proceeds along two pathways:

One set of reactions, Process 1, consumes Br$^-$, and when [Br$^-$] falls below a critical level, then

Process 2 takes over. But Process 2, during which $Ce^{+3} \rightarrow Ce^{+4}$, produces more Br$^-$. Then the Ce^{+4} oxidizes malonic acid to produce the reactants of Process 1. Thus the reactions oscillate back and forth (with color changes) until all the malonic acid is consumed.

5.4 Nerve Conduction

A 0.1 mA current acting for 0.1 ms delivers a charge of 10^{-8} coulomb. If this charge is delivered to a capacitor $C = 10^{-6}$ farads, then the capacitor has a voltage of $V = Q/C = 10$ mV. If we measure current in mA, time in ms, and capacitance in mF, then

$$V = \frac{(0.1\,\text{mA})\ (0.1\,\text{ms})}{.001\,\text{mF}} = 10\,\text{mV}.$$

Richard Feynman in his Lectures on Physics Volume 1, Chapter 3 compares nerve conduction to the knocking-over of a line of dominos. An initial tap converts the stored energy into a travelling wave.

5.6 Space Shuttle Launch

The Thrust T of the Shuttle engines is actually throttled back toward the end of the first-and-second stage burns, to keep the acceleration within strict (structural) limits. This can be modeled in the Diff Eqs by making T(t) a user-defined function of time.

Problem 5-5 Flight to the Stars

The origins of this problem, accelerated relativistic flight, are somewhat uncertain. Sebastian von Hoerner in his 1962 paper in Science 137 , 18-23 "The General Limits of Space Travel" derives the relevant equations in analytic form. However there is a very interesting book The Realities of SpaceTravel (selected papers of the British Interplanetary Society, McGraw-Hill, 1957) that would indicate that J. Ackeret and L. R. Shepherd had worked out the solutions to this problem as early as 1952.

A Final Note

This is the end of the text, however I hope you will find, as I have, that the computer with modern mathematical software is a wonderful tool for investigating Physics, and physics-related problems. If you have found new applications for mathematics applied to the physical sciences, please send them along.

Ideas for new problems and their solutions are always welcome.

stevevw@u.washington.edu

svanwyk@oc.ctc.edu

References

Chapter 1

1-1 David and Judith Goodstein, Ed. (1996) <u>Feynman's Lost Lecture</u> [Norton]

1-2 Jerry Marion (1970) <u>Classical Dynamics</u> Chapter 8 [Academic Press]

1-3 James van Allen (2003) *"Gravitational Assist"* Amer. J. Phys. <u>71</u> , 448

1-4 Robert Adair (2002) <u>The Physics of Baseball</u> [Harper-Collins, 3ed]

1-5 PSSC Physics (any edition) the Physical Science Study Committee

1-6 Dan Poynter, Mike Turoff (2003) <u>Parachuting, the Skydivers Handbook</u> [Para Publishing, 9ed]

Chapter 2

2-1 Richard Feynman (1964) <u>The Feynman Lectures on Physics</u> Volume 1, Chapter 23 [Addison–Wesley]

2-2 Robert Vierck (1979) <u>Vibration Analysis</u> [International Textbooks, 2ed]

2-3 Ray Wylie, Louis Barrett (1995) <u>Advanced Engineering Mathematics</u> Chapter 8 [McGraw–Hill, 6ed]

2-4 Brice Carnahan, Herbert Luther, James Wilkes (1969) <u>Applied Numerical Methods</u> [Wiley]

2-5 Sir Arthur Eddington (1926) <u>The Internal Constitution of the Stars</u> [Cambridge University Press]

2-5 Robert Christy (1964) *"Calculation of Stellar Pulsation"* Reviews of Modern Physics <u>36</u> , 555

2-5 Bradley Carroll, Dale Ostlie (1996) <u>Introduction to Modern Astrophysics</u> Chapter 14 [Addison–Wesley]

Chapter 3

3-2 JackWilmore, David Costill (2004) <u>Physiology of Sport and Exercise</u> [Human Kinetics Publications]

3-2 Henry Bent (1978) *"Energy and Exercise"* J. Chemical Education <u>55</u> , 797

3-2 William Pritchard (1993) *"Mathematical Models of Running"* SIAM Review <u>35</u>, 359

3-3 http://www.JohnstonsArchive.net/astro/index.html
-- For data and references on the relativistic binary pulsar PSR 1913+16

3-4 Romas Mitalas(1980) *"Supernova in Binary Systems"* Am. J Phys <u>48</u> , 226

3-5 Jack Holman (1997) <u>Heat Transfer</u> Chapter 10 [McGraw–Hill, 8ed]

Chapter 4

4-1 Horatio Carslaw, John Jaeger(1959) <u>Heat Conduction in Solids</u> [Oxford]

4-3 Frank Incropera, David DeWitt (2002) <u>Introduction to Heat Transfer</u>
     [Wiley]

4-4 David Derbes (1996) *"Feynman's Derivation of Schrodinger's Equation"*
     American Journal of Physics <u>64</u> , 881

4-4 Stephen Gasiorowicz (1996) <u>Quantum Physics</u> Chapter 5 [Wiley, 2ed]

4-5 Leonard Schiff (1968) <u>Quantum Mechanics</u> Chapter 13 [McGraw-Hill]

4-5 Kenneth Krane (1988) <u>Introductory Nuclear Physics</u> [Wiley]

4-5 Richard Robinett (1997) <u>Quantum Mechanics with Visualized Examples</u>
     [Oxford]

Chapter 5

5-1 Peter Milonni, Joseph Eberly (1988) <u>Lasers</u> [Wiley]

5-1 WilliamWagner,Bela Lengyel (1963)*"Evolution of Giant Pulse in a Laser"*
     Journal of Applied Physics <u>34</u>, 2040

5-2 Bassam Shakhashiri (1985) <u>Chemical Demonstrations</u> Volume 2,
     Chapter 7 [U. Wisconsin Press]

5-2 John Tyson (1976) *"The Belousov-Zhabotinski Reaction"* Chapter 2
     in <u>Lecture Notes on Biomathematics</u> Volume 10 [Springer]

5-3 George Gause (1934) <u>The Struggle for Existence</u> [Dover reprint 2003]

5-3 Narendra Goel, Samaresh Maitra, Elliot Montroll (1971) *"On the Volterra
     Model of Interacting Populations"* Reviews of Modern Physics <u>43</u> , 231

5-4 Alan Hodgkin, Andrew Huxley (1952) *"A Quantitative Description of
     Membrane Current and Excitation in Nerve"* Journal of Physiology <u>117</u> , 500

5-4 Russell Hobbie (1997) <u>Intermediate Physics for Medicine and Biology</u>
     [AIP Press]

5-5 J Rand McNally (1982) *"The Physics of Fusion Fuel Cycles"*
     Nuclear Technology/Fusion <u>2</u>, 1

5-6 Kerry Joels, David Larkin, Greg Kennedy (1988) <u>Space Shuttle Operator's
     Manual</u> [Ballantine]

Constants

M_{sun} = 1.989×10^{30} kg $\qquad$ R_{sun} = 7×10^5 km

M_{Earth} = 5.98×10^{24} kg $\qquad$ R_{Earth} = 6380 km

M_{Moon} = 7.35×10^{22} kg $\qquad$ R_{Moon} = 1740 km

$M_{Jupiter}$ = 1.90×10^{27} kg $\qquad$ $R_{Jupiter}$ = 71000 km

G = Gravitational constant = $6.673 \times 10^{-11} \dfrac{N \cdot m^2}{kg^2}$ = $6.673 \times 10^{-11} \dfrac{m^3}{kg \cdot s^2}$

with distance in km, $\quad G = 6.673 \times 10^{-20} \dfrac{km^3}{kg \cdot s^2}$

h = 6.626075×10^{-34} J-s

$\hbar$ = 1.054572×10^{-34} J-s $\qquad$ 1 eV = 1.602177×10^{-19} J

m_p = 1.672×10^{-27} kg

m_n = 1.675×10^{-27} kg $\qquad$ q_e = 1.602177×10^{-19} C

m_e = 9.109389×10^{-31} kg

c = 3.00×10^8 m/s $\qquad$ σ = 5.67×10^{-8} W/m^2-K^4

$\qquad\qquad\qquad\qquad\qquad$ k = 1.3807×10^{-23} J/ K

1 AU = 149.6×10^6 km = 149.6×10^9 m

1 yr = 3.1557×10^7 s $\qquad$ ϵ_o = 8.854188×10^{-12} C^2/N·m^2

1 g = 9.81 m/s^2 = 32.16 ft/s^2

Index